The
PATIENT
as
CEO

How Technology Empowers
the Healthcare Consumer

ROBIN FARMANFARMAIAN

THE PATIENT AS CEO

How Technology Empowers

the Healthcare Consumer

ISBN 978-1-61961-376-8

LIONCREST
PUBLISHING

I'm dedicating this book to my mom
Dr. Wendy Tinklepaugh Soloway
(1942–2008), my role model.

Praise for The Patient as CEO

"Health and medicine have become an information technology and are therefore progressing at an exponential pace. Many physicians are unaware of the many new treatments that are now on the edge of clinical practice. As a result, it is increasingly important for patients to take charge of their own health by tracking and understanding these new developments. Robin Farmanfarmaian has written an excellent book to help guide patients through this emerging world of healthcare. Especially moving and instructive is her own courageous experience as a patient, which she eloquently describes."

— RAY KURZWEIL, INVENTOR, AUTHOR, AND FUTURIST

"We are experiencing a medical revolution where exponential technologies will make each of us 'the CEO of our own health.' Robin personifies that message—she saved her own life by taking charge of her own treatment. A must read for patients to understand the current medical revolution driven by accelerating technologies."

— PETER H. DIAMANDIS, MD, CO-FOUNDER, EXEC. CHAIRMAN, SINGULARITY UNIVERSITY, CHAIRMAN, XPRIZE FOUNDATION

"The Patient as CEO may be the biggest transformation in the doctor-patient relationship since Hippocrates. Robin Farmanfarmaian offers a backstage pass to the accelerating innovations on the leading edge of medicine."

— ALAN GREENE, MD

"Robin's story reinforces my belief in 'INsight'—that is, seeing INTO our physiological processes and understanding them in ways never before possible, will change medicine forever."

— VINT CERF, INTERNET PIONEER

"Robin Farmanfarmaian provides a fascinating look into the future of medicine in her book, The Patient as CEO. Starting with her own agonizing journey, she writes about how the future is happening faster than we think—with a wide range of technologies converging and making science fiction possible."

— VIVEK WADHWA, TECHNOLOGY ENTREPRENEUR AND ACADEMIC

"Extensively researched, and filled with personal experience, Robin makes complicated medical concepts clear in an engaging conversational tone. An excellent guide for anyone facing taking control of their own healthcare and wanting to understand the potential of future healthcare technologies."

— CATHERINE MOHR, MD MSME

"The Patient as CEO is today's personal handbook to health and wellness. It reads with the same pace and vigor as the very technology that is transforming our lives. Robin Farmanfarmaian take us on a unique journey that informs and empowers us all to take control our our health and become the chief executive officer of our lives and our destiny. If you have a breath or a pulse, this books is a must-read and

your owners manual to help navigate the essential role of technology in keeping those beats, breaths and smiles alive and well!"

— JOHN NOSTA, FOUNDER, NOSTALAB, MEMBER, GOOGLE HEALTH ADVISORY BOARD

"Most patients are passive. Not Robin. Combining extreme optimism and extreme technology her relentless drive attempts to rip up and redesign the existing medical system into a world where each patient is gene sequenced, measured, targeted by a specific therapy. If half of what is described in this book comes to pass... our kids will live far longer."

— JUAN ENRIQUEZ, CO-AUTHOR EVOLVING OURSELVES, MANAGING DIRECTOR, EXCEL VENTURE MANAGEMENT

"Simply the best written clearest snapshot of the medical technology pipeline. Read it today, place your own bets on what will win, and read it again next year to find out. The "who" will win is obvious, it is you the patient."

— KEVIN R STONE MD, ORTHOPAEDIC SURGEON, THE STONE CLINIC, SAN FRANCISCO

"Robin Farmanfarmaian delivers a pointed, lightning-quick tour through a dizzying array of possibilities rapidly emerging in medical and health technologies. From artificial intelligence to repurposed drugs, and robots to playing neural games to treat diseases of the brain, these discoveries, gizmos and gadgets are fusing together to radically change how we heal and who we are as humans. I can think of no better guide through this blizzard of emerging medtech than Robin, a font of energy whose own experiences with what can go right and wrong in medicine exemplifies perhaps the most crucial transformation of all—the rise of the patient and the medical

consumer as a major contributor to their own health and well being. This is a brave book and a romp through the power of what is possible."

— DAVID EWING DUNCAN, AUTHOR OF THE
BESTSELLING EXPERIMENTAL MAN

"This book is about so much more than pinpointing the true healthcare trendsetters and emerging technologies that will give readers (patients, caregivers, students, researchers, and physicians alike) a crystal ball to forecast the coming future of medicine. This book is about the central idea that patients everywhere on this planet are sensing: "HOPE". Robin reveals, in easy to follow language, the hope that rapid advances in fields as diverse as artificial intelligence to biometric sensors to nanotechnology to autonomous drones will forever change the landscape of healthcare to one with the patient in the center of it all. There is hope that the patient will be empowered to take on their deserved chief executive officer role to navigate their own health care strategy in unprecedented ways. Patients and their families everywhere should read this book as the new playbook for personalized medicine."

— DR. KETAN PATEL, CEO, NEXT
HEALTHCARE TECHNOLOGIES, INC.

"The idea of patients taking control of their own health might sound radical to some; yet that is the future. In The Patient as CEO, Robin Farmanfarmaian eloquently outlines how this might happen. Human healthspans are set to grow and one of the drivers will be a transfer of power from the current system to individuals and their families. This book is a must read!"

— SONIA ARRISON, AUTHOR, 100 PLUS: HOW THE COMING
AGE OF LONGEVITY WILL CHANGE EVERYTHING, FROM
CAREERS AND RELATIONSHIPS TO FAMILY AND FAITH

"*The brilliance of this book is the effortlessness with which Robin Farmanfarmaian draws the reader through her personal story into the world of the transfer of power from our fragmented Health Care system to Patients. She highlights the role of Exponential Technologies that are helping to re-establish the connection between patients and appropriate and timely care that they deserve. In addition to patients, this book is a must read for everyone in the health care field so they can be part of the evolution to a better future.*"

— RAMESH KUMAR MD, THE PATIENT DOCTOR

"*The convergence of accelerating technology will disrupt 40% of the Fortune 500 by 2020. Healthcare is one of the industries that will be impacted the most. Robin Farmanfarmaian lays out a realistic scenario of what the world will be like for patients of the near future.*"

— KIAN GOHAR, EXECUTIVE DIRECTOR, INNOVATION
PARTNERSHIP PROGRAM, XPRIZE

"*Drawing on her own journey through the healthcare system, Robin provides us all with a powerful call to action for all of us to shift from passive patients to empowered CEO's of our health and wellness. Along the way, she offers a compelling and panoramic view of the big shift that is occurring in the healthcare industry, shaped by exponential digital technology. Read this book—your life may depend upon it!*"

— JOHN HAGEL, CHAIRMAN OF THE DELOITTE
CENTER FOR THE EDGE AND PROLIFIC AUTHOR

"*Technology is driving massive innovation right now, your success and impact is only limited by your imagination and ability to think BIG. Robin imagines a truly personalized future for the patient—a great overview of the accelerating technologies enabling this future to be the reality."*

— NAVEEN JAIN, FOUNDER, MOON EXPRESS, WORLD INNOVATION INSTITUTE, INOME

"Knowledge is power. Technology democratizes information and empowers us to make better choices about our health. If you wonder why technology is disrupting healthcare and how it can be used in healing ways, read Robin Farmanfarmaian's book today. She speaks from experience, seamlessly weaving her own heroic medical history as a chronic disease patient with her seminal role at Singularity University into her vision of how we can co-create real health care, not just sick care."

— DEAN ORNISH, M.D., FOUNDER & PRESIDENT, PREVENTIVE MEDICINE RESEARCH INSTITUTE, CLINICAL PROFESSOR OF MEDICINE, UCSF, AUTHOR, THE SPECTRUM

Contents

ABOUT THE AUTHOR

ROBIN FARMANFARMAIAN BELIEVES THAT TECHNOLOGY can empower patients and make a positive impact in the health and medical field. This position drives her to provide education and resources to leaders, entrepreneurs, physicians, healthcare professionals, and innovators to positively impact medicine and healthcare.

Farmanfarmaian focuses on the future of integrated medicine, the changing role of patients in healthcare decision-making, and how technology will change the way we experience and interact with medical facilities and physicians. She is interested in big data, wearable technology, 3-D printing, and access to and the usability of personal healthcare information.

Farmanfarmaian is a professional speaker on technology and medicine for many conferences and companies, including Exponential Medicine, Singularity University, Connected

Health Symposium, Boston Scientific, Differential Medicine, Medscape CME Videos, Health 2.0, LSA Innovator Summit, Wharton, and the Kellogg School of Management at Northwestern University, among many others. She is adjunct faculty for Singularity University and a contributing writer to *Wired, Forbes, Huffington Post, MedGadget, Becker's Hospital, Fierce-HealthIT,* and she has been published in a variety of other online publications.

Currently, Farmanfarmaian is working with a few early stage start-ups: COO for Arc Fusion Programs, on the fusion of health, science, and IT; cofounder, Board of Directors, and former Executive Director for the Organ Preservation Alliance, catalyzing breakthroughs in transplants, organ banking, cryopreservation, and tissue engineering; VP of Business Development with Invicta Medical, a device company for acute care/post-anesthesia and sleep apnea, poised to impact hundreds of millions of patients; and President of i4j ECO, a summit to disrupt unemployment through innovation to create jobs and meaningful work for everyone.

Other work experience includes being one of the founders of the Exponential Medicine Conference—a conference for physicians and healthcare executives on the next five to ten years in medicine and how technology will be impacting and disrupting healthcare. She was the cofounder and Chief Business Development Officer for MORFIT, a fitness platform. She was also the Vice President of Strategic Relations for Singularity University, which trains executives and leaders to drive global change and to stay competitive through the use of the convergence of accelerating technology.

She serves on the advisory boards of many early stage start-up tech companies and organizations such as sxsw V2Ventures, AARP, Doctella, and the Kairos Society, among others. She mentors women in entrepreneurship, technology, and healthcare and holds a series of dinners for women in technology.

To book Robin for speaking engagements and consulting, please email her at: ThePatientCEO@gmail.com

ACKNOWLEDGMENTS

I HAVE BEEN LUCKY ENOUGH TO HAVE AN ABUNDANCE OF supportive family and friends. Huge thank-yous to my amazing dad Nick Soloway, who went through so much with me, and to my immediate family—Chris Soloway, Kate Soloway, Nicholas Soloway, Kay Ransdell, and Karena Akhavein. I am the luckiest person in the world.

I also want to thank other uber supportive friends and Robin-cheerleaders: Silvia Console Battilana, Taylor Milsal, Valentina Morigi, Jim Kwik, Nathana O'Brien, David Ewing Duncan, Cheryl and Alan Greene, John Hagel, Joe Polish, Vivek Wadhwa, and the incredible teams at INVICTA Medical, the Organ Preservation Alliance, i4j, and Arc Fusion.

And of course, Ray Kurzweil, Peter Diamandis, and the faculty at SU for teaching me so much.

A huge thank-you to all the people I mention in this book. Most are friends, most spent significant time talking with me about medical technology, and all are making a major impact in medicine. I've been fortunate to have a lot of them in my life for years now. This book wouldn't have been possible without their contributions to medicine and to my life.

And thank-you to all the health care professionals who helped me over the years.

Enjoy!

Introduction

EMPOWERING AND ENABLING PATIENTS WITH FAST-MOVING TECHNOLOGY

A CONVERGENCE OF TECHNOLOGIES IS HAPPENING IN MEDicine. We're entering a perfect storm of technological advancements that are enabling the era of the patient. We're now in an era of not only patient-focused care but also patient-directed medicine.

This book is written for anyone who will ever be a patient, written by a patient studying and working on the cutting edge of med and biotech. It is for anyone who is interested in the convergence of accelerating technology in medicine and healthcare. Specifically, this book is for anyone curious about how technology will impact patients.

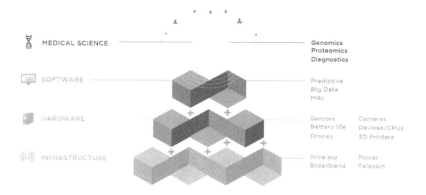

It is for anyone who wants to understand that for a patient, technology is hope.

What's my story? I'm a serial entrepreneur. I've worked on more than ten early stage start-up companies, all dealing with cutting edge technology, mostly in medicine and biotech. My driving force comes from my personal experience. Back when I was a teenager, I was misdiagnosed with an autoimmune disease that resulted in 43 hospitalizations and six major surgeries.

When you're facing surgery, especially when you're just a teenager, you look for the best doctors and hospital systems. You get a second opinion, a third opinion, a tenth opinion...

But none of my doctors looked at me and said, "You know what, Robin, let's hold off on the surgery because you're so young. We can afford to wait for some innovation in treatments." Nobody looked at me and said, "Technology is hope."

But I've learned that technology *is* hope! In fact, at the time of my initial surgeries, laparoscopic surgery had already been invented. It was just so new that it wasn't in widespread use. If I had waited just a few more years for my surgeries, laparoscopic techniques would have reduced five of my surgeries to just one, with an incision measured in millimeters instead of inches. Huge difference in recovery, risk, and my life.

Had I waited for just a few more years after that, I could have been treated with an entirely new class of drugs called anti-tumor necrosis factor biologics. Had I known this game changing drug was only a few years away, I could have waited for it, and it would have eradicated the need for me to ever have surgery at all. Tumor necrosis factor, also known as TNF alpha, was discovered by a team that included my hero, Dr. Bob Hariri. TNF alpha can put many different diseases, such as rheumatoid arthritis and Crohn's disease, into remission. Had I, or my physicians, known what disruptive technologies would be impacting medicine in just the next few years, I may not have had three organs removed.

About ten years into my misdiagnosis, after my main surgeries, my doctors had me on high-dose opiates and a maintenance dose of steroids. I was told I'd probably be on both types of medications for the rest of my life. There was no other treatment option, as the surgery had supposedly "cured" me. They even started talking about surgically implanting a morphine pump in my spine to treat my extreme pain, which they thought was from so much abdominal surgery. Living on high-dose pain killers and steroids for the rest of my life? I was 26 years old! No way.

I thought they were wrong. The pain wasn't from the surgeries. Something else needed treatment, and I was going to figure out what it was. So one day, I decided to take full control of my own medical care, and at the age of 26, I fired all of my doctors. All of them.

I put together a whole new medical team—one that worked collaboratively with me. They gave me my lab results, explained them to me, and allowed me to decide on tests and treatment plans. They helped me get off *all* the medication. Soon afterward, I was finally diagnosed correctly and put on the anti-TNF biologic Remicade. My life changed overnight, literally. The IV dose of Remicade took only 24 hours to kick in. After 13 years of extreme pain, I was suddenly a different person. Not only did becoming the CEO of my own healthcare team potentially save my life but it also gave me *back* my life.

A patient writing a technology book?

I've sat down and spoken with almost all of the people mentioned in this book. We talk tech, trends, science, entrepreneurship, and healthcare. I attended the majority of presentations at Singularity University for over three years. I took dives into major areas of accelerating technology, including robotics, computing systems, biology, nanotechnology, artificial intelligence, future studies, space, global challenges, incentivized prizes, and entrepreneurship. I took very deep dives into medicine. I've spoken at, organized, or attended many other educational channels in medicine, biology, and technology, including Arc Fusion Programs, Health 2.0, TEDMED, TED, Connected Health Symposium, Differential Medicine, and workshops on tissue engineering and organ banking for the White House and DARPA with the Organ Preservation Alliance, among many others. I've listened to literally thousands of hours of content on technology, medicine, and science outside of my formal education. I've taken that knowledge and broken it down for you, as told through the eyes of a patient, to show just how rapidly medicine is changing, empowering, and enabling the patient to be a key decision maker.

A Futuristic Scenario:

With virtual reality, point of care diagnostics, AI, robots, and sensors that track your every biometric activity, how much would you need to leave home if you had a chronic condition? Work from home. Play from home. Education in a VR classroom, complete with tactile sensing. Robot companions with artificial intelligence. Your body and environment more finely tracked, recorded, and analyzed than a plane or rocket ship with subcutaneous sensors, sensors inside blood vessels, in your toilet, and your house. Brain computer interfaces communicating externally, controlling external robots, robotic limbs, and computers, and used to communicate without talking. Knowing exactly how many calories you need and the calorie counts of all the foods you eat, when to exercise, and exactly what to eat based on your body's needs, tailored to your favorite tastes. Medication tweaked with every dose, 3-D printing each unique, custom tailored dose to you in your own home. Sensors detecting if you took the drug on schedule, if you overdosed, if you're having a reaction to the drug, and if you need a new or different drug. Deliveries of everything on demand by autonomous drone or autonomous car. When you need more milk or are deficient in a nutrient, your refrigerator or house will order it instantly, and it will seamlessly appear in your fridge, delivered by an autonomous car and handed to your autonomous robot home companion to put away.

That futuristic scenario is becoming a reality. Imagine the impact this will have on patients. It opens up a world of independence, a world of no longer needing to spend their lives in waiting rooms, a world of no longer needing to rely on a small army of other people to help you manage your life and disease, and a world where all outcomes are improved.

DATA AND EMPOWERMENT

Access to data about your own health is really empowering to you as a patient. It helps you look at your body as an operating system. We are moving into a world where you will be able to collect accurate data that seamlessly tells you everything about your health; sleep, activity, continuous blood labs, and vital signs. That's very empowering because you can gather this information on your own—no need for a finger prick and blood analysis at a lab or the doctor's office.

When I was being tapered off steroids, every single Monday morning for a year and a half, I had to go to a lab at 8 a.m. and get blood drawn to check my cortisol levels. It was super inconvenient to drive to the hospital, park, go to the lab, wait, have the blood drawn, and then get home, which took at least an hour, plus my health insurance co-pay and the fact that I was exposed to potential infectious agents while coming off a medication that suppressed my immune system. It was dangerous!

What if there was a simple sensor that could tell your cortisol level whenever you wanted, wherever you were, and without a blood draw? Such a device would have saved me maybe 75 to 100 hours of my life, to say nothing of the pain from being stuck with a needle, the risks of the blood draw, and the cost to my insurer. For patients with compromised immune systems, just going to the lab is a risk. Hospitals are full of sick people, and you could easily catch something. What if you could eliminate that risk by never having to go to the lab? Inexpensive, accurate monitoring devices will be game changers for huge numbers of patients.

Technology that gives access to information and patient communities is equally empowering. When I was growing up, my only access to medical information, beyond my 15 minute doctor visits, was my mom's six-inch thick *Physician's Desk Reference,* which isn't meant for nonmedical readers (she was a pediatrician). Now, with "Dr. Google," we have access to vast amounts of information. We're no longer passive victims of a disease. Now we're empowered by knowledge. You can educate yourself about a medical issue in ways you couldn't even ten years ago. Online patient communities let you educate yourself faster by talking with people who share your medical issues and who can provide valuable, practical, and emotional support.

THE EMERGING ERA OF THE PATIENT: A PARADIGM SHIFT

Until recently, the patient didn't feel like the customer in the medical system; the patient often felt like an afterthought. Patients haven't been directing their own care, and they're not always in on the planning. Let's start looking at and rearranging medicine around the person we're supposed to be helping. Patients should be the ones in control. Instead of patients being outside the inner circle that makes medical decisions, they should be dead center in that diagram. They, or their caregivers, should be the CEOs of their own healthcare team.

Healthcare can now be two-way, where patients are more informed and empowered regarding their health. They can become more involved in their own healthcare by researching their conditions, tracking their symptoms, and measuring their progress. Physicians, in turn, can save time, see more patients, and engage in more productive, insightful one-on-one discussions.

The new era of doctor-patient relationships means that patients are no longer on the sidelines when it comes to their health. In fact, the role of the patient has shifted from benchwarmer to star player. Visiting the doctor is like checking in with a coach; patients come to their appointment, review their progress, listen to recommendations, go home, and try to improve.

All the emerging technologies I'll discuss in this book are helping that happen. When we start bringing diagnostics out of the hands of the hospital and the doctor, we have the opportunity to put it right into the patient's hands. Sensor technologies and wearable devices are becoming realities, as companies like Scanadu are showing. Scanadu makes a suite of consumer medical device products that empower patients by giving them access to data that they never had before, such as tracking their blood pressure. They gain insight into their own health and provide useful information to their physicians.

Your car is full of sensors, as is an airplane. The sensors tell you about the car or plane's operating system and record data that can be accessed to diagnose a problem. Look at your body in the same way you would a car or a plane. You are an operating system (software) combined with hardware that can be broken down into components with data. The hardware is the part you can see and touch—your organs, your skin, and your bones. The software is how your genes and brain and neurology all work in conjunction to keep your hardware healthy, functioning, and moving. Breaking the body down into an operating system makes you realize that everything inside it is treatable in some way. That's what we're striving for now, finding ways to monitor and tweak the operating system of your body, instead of modifying the hardware by, say, removing an organ.

Artificial intelligence (AI) is also going to be very empowering for patients. AI is machine intelligence—computers that can perceive, reason, and imagine much as humans do. In other words, AI refers to a really, really smart computer that can think for itself, understand normal spoken language, and communicate with you seamlessly.

Most of us are already using AI on a daily basis and have for years now! When you ask Siri something on your smart phone, you're using AI. You'll be using it a lot more in the future. AI will complement, supplement, or even replace doctors for much basic medical care. Why drag yourself to the doctor's office when you feel sick when you could stay in bed and let an AI "doctor" diagnose the problem and prescribe the remedy? Why wait days for an appointment when you can "see" the AI doctor immediately without having to take time off from work? Talking to the AI doctor could eliminate a lot of time-consuming office visits. And the AI doctor never sleeps, so your visit can be any time of day or night.

A decade from now, we will be fully equipped with personal

sensors in the same way a plane is equipped with sensors. They will be silently collecting data from every single aspect of your body, continuously and accurately, without bias, and giving it to you. When you start to get sick, the sensors will know before you do. Before you start to feel the first symptom, your sensors will see that some readings are off. When markers start to converge, you'll be notified. If you're starting to get a little dehydrated, your sensors will tell you and not only suggest that you have something to drink but also remember what your preferred drinks are. In the future, your sensors will be hooked up to your "smart house" as part of the IoT (Internet of Things). The sensors could automatically deploy a robot to bring you your favorite drink and automatically replenish your drink supply by placing an order on a crowdsourced grocery store delivery service like Instacart.

If your blood sugar is dropping too low, you'll be told so that you can take action before the symptoms get severe. That's for people even without diabetes or hypoglycemia, so they can stay in perfect working condition. For people with diseases like diabetes who need constant monitoring to stay alive, this is a game changer, and it means no longer needing to remember to prick your finger (especially for children!), not to mention going through the pain. If the situation gets to the point where it looks like you need medical attention, your sensor system will immediately alert your doctor. You won't even have to make a phone call; they're going to call you. It might be the AI doctor calling to tell you that you're not feeling well because you forgot to take your blood pressure medication or that your sensor readings say you need to head to the ER right now.

According to a McKinsey study, by the year 2025, the health part of the IoT will be $1.6 trillion globally. McKinsey defines the health set of the IoT as "devices attached to or inside the human body...people will use to guide their actions and decisions."

ACCELERATING AND EMERGING TECHNOLOGIES

I've heard people say, "Sure, all this sounds great, but it's way too complex and expensive to be widely used." Remember that back in the early 1980s, VCRs were considered too complicated to hook up and use, plus they were very expensive. Then, simultaneously, everyone figured out how to use them (except the flashing clocks), and the price became very, very affordable to the point where everyone had a VCR and used it all the time. Today? The technology is so outdated you can't even give them away. The same thing happened with DVDs and game consoles. Just look at your entertainment technology now to see how that trend can be applied to medical devices and medical diagnostic tools. We've seen this over and over throughout tech history. When exponential growth starts, it moves very rapidly.

We're going to start to see very rapid acceleration of all the technologies I'll discuss in this book. A fantastic example of accelerating technology is the iPhone. Once, it didn't exist. We had landlines, then came the massive game changer of cellular technology. Phones were getting smaller and cheaper and coverage was improving. And then, bam! The iPhone hit the market in 2007, and it literally changed everything around the world very, very quickly. Can you imagine your life now without it even for a day? An hour?

In the medical world, in the past couple of decades accelerating technology has given us small, accurate glucose meters to replace inaccurate, inconvenient urine test strips for people with diabetes. Laparoscopic surgery has made complex surgery easier and safer for patients, using tiny incisions with much shorter recovery times. MRIS give us unprecedented accurate views inside the body. Today, we can do smart phone physicals using an array of small devices that attach to your phone.

THE PATIENT OF THE VERY NEAR FUTURE

In the very near future, being a patient will be remarkably different. You'll collect your own health data using a device like Dave Albert's AliveCor heart monitor iPhone case to do your own EKG. You'll use an app on your smart phone that analyzes your cough. You'll send that data to the cloud where a physician will read it from anywhere in the world. If you need further discussion or treatment, instead of an office visit, you might use telemedicine or an AI physician. And if you need medicine, a drone will deliver it to your door.

As the science fiction writer William Gibson frequently says, "The future is already here—it's just not very evenly distributed." If you live long enough, you have a 100 percent chance of getting sick, potentially with something serious, terrifying, and painful. Accelerating technology in medicine will help you put that off as long as possible by helping you prevent chronic disease or manage your disease more effectively. If you do get sick, it will be able to assuage some of your fear with access to information, both from your own body's software as well as about your disease. It will help you live your life to the fullest and be as healthy as you can possibly be, even if you have a chronic disease or disability. Being a patient or being disabled is no longer a disadvantage. Technology is more than an equalizer; it can be an upgrade!

In this book, I mention lots of companies that are doing innovative things in healthcare. Some are huge corporations that are pouring vast amounts of money into technology development. Others are start-ups, some still in stealth mode and some in mega growth mode. I've spoken with researchers and executives at many of these companies, and I'm deeply impressed by their intelligence and vision. Will every idea in this book still be viable even a few years from now? Will every company be successful? Probably not. But with every success and every failure, the era of patient-directed medicine gets closer.

Chapter One

WEARABLE TECHNOLOGY AND SENSORS

WEARABLE TECHNOLOGY IS DEFINITELY ON AN ACCELER-ating trend. Activity tracking devices such as Fitbit, Nike Fuelband, the vast array of smart watches, and even your iPhone use three-dimensional accelerometers to measure how much you move. All the major tech companies are coming out with their own version of smart devices like the Apple iWatch, which can be used to track your biometric activity with your smart phone. Lots of different apps to track your activity level and some basic biometrics are now available on your smart phone. Unfortunately, not all of them are that accurate. Most devices are still clunky, though some companies are now taking style into consideration, like Apple's partnership with the couture designer Hermès or wearables pioneer and forward-thinking visionary Sonny Vu, founder and CEO of Misfit Wearables (recently acquired by the Fossil Group for $260M), which makes the Shine activity tracker. Sonny built the Shine with fashion as a cornerstone from inception, the first company to do so successfully. Misfit offers a range of sleek accessories that incorporate the tracker and has now partnered with Swarovski to create jewelry that can hold the tracker.

Wearable tech no longer just passively records. You can now interact with it, or even better, just ignore it. As a patient, if you can ignore your monitoring technology while it quietly works in the background, then your engagement with it is going to go up to 100 percent. For example, we're starting to see EEG stickers that can monitor brain activity by tracking brain waves. A very lightweight, miniaturized EEG system can now be placed behind the ear and worn continuously for up to two weeks. The patient doesn't have to do anything and barely knows the monitor is even there—unless it detects something abnormal and sends an alert.

Many start-up companies, like Medella Health, are working on contact lens sensors that can continuously track blood glucose levels. Google has partnered with Novartis to create their version. Both Google and Novartis have patents on how the lenses are powered: radio waves from a wearable external device that could be incorporated into jewelry, hats, or other pieces of clothing. In the future, we'll be able to use contact lenses to monitor other things in your body, like the level of a drug.

For people with diabetes, continuous monitoring sensors will have a huge impact. Multiple finger pricks every day to check your blood glucose will soon be a thing of the past. Dangerous low blood sugar episodes will be avoided because your sensor will alert you before you drop too low. When the discomfort and expense of finger pricks goes away (currently a single test strip for a glucose meter costs about a dollar), compliance with diabetes treatment is going to go way up. We're going to start to see fewer complications you can get from diabetes, like foot ulcers, amputations, and vision loss.

At iHealth, Uwe Diegel has created a company that designs and manufactures innovative, consumer friendly, mobile personal healthcare products that connect to the cloud. You can track your vital signs on your smart phone or computer and share the results with caregivers, family members, and your healthcare team. The devices are comfortable and easy to use with free apps. You can use iHealth monitors to track your blood pressure and monitor blood glucose, pulse and oxygen saturation, activity, and sleep. When you look at each data point separately, you learn more about your health. But when you look at all the data points in aggregate, you learn a lot more—maybe even enough to keep you out of the hospital.

Wearable technology is enabling large, data-driven projects that have huge potential for understanding all the factors that affect our health. The Google Baseline study, begun in 2014, will use a combination of genetic testing and digital health sensors to collect baseline data on healthy people. The idea is to establish genetic biomarkers relating to "how [patients] metabolize food, nutrients and drugs, how fast their hearts beat under stress and how chemical reactions change the behavior of their genes."

Separate from the Baseline Study, the Google X life sciences group is working with Biogen on a project to study outside factors that could contribute to multiple sclerosis. Google and Biogen will use sensors, software, and data analysis tools to collect and analyze data from people who have MS. The companies aim to explore why this disease progresses differently in each patient. This sort of research will eventually help MS patients get treatment that is tailored precisely to their symptoms.

The demand for wearable monitors is already strong. According to a worldwide consumer survey published in March 2015, 70 percent of US consumers want a health wearable or app that monitors every aspect of their health. In another recent survey, 8 percent of consumers already own at least one wearable fitness monitor and 6 percent own a wearable health monitor. The potential market is huge, especially because health insurers are increasingly willing to pay for the devices. In 2014, every major vendor made or announced a wearable device: Moto 360, Samsung Gear, Microsoft Band, Apple Watch, and others. According to IDC research, 19.2 million shipments in 2015 will build to a global market of 111.9 million units by 2018. The global wearable market is growing at a 43.4 percent compound annual growth rate, forecasted to reach $30.2 billion in 2018.

Wearable technology is an industry in its infancy. Right now, fitness tracking with wearable technology is fun and educational. We're soon going to see the same kind of technology mature, get much more accurate, and do much, much more. People with chronic conditions will really benefit from being able to see the bigger picture of their own health. We're going to see more wearables become HIPAA compliant and get FDA approval. We're going to see sensors that fit against the skin (epidermal) or under the skin (subcutaneous). We're going to see more permanent sensors that fit inside your body, including inside blood vessels.

TECH IN YOUR TROUSERS

Could your bra be your next doctor? A company called Ohmatex creates smart textiles that have sensors woven in to them. Smart textiles conform to your body and let you move naturally. Ohmatex makes socks that monitor for edema or swelling in the feet from fluid retention—a warning sign of worsening heart failure. A company called Sensoria is developing socks with sensors in them to measure gait for neurodegenerative diseases. Sensors are now widely used in firefighting gear to detect overheating and sound an alarm if a firefighter is in danger.

A number of companies, including Ralph Lauren Polo (partnered with OMsignal) and Intel, now make smart shirts that are used to monitor athletic performance—two brands you never would have thought would be direct competitors with each other (and that I never thought I would use in the same sentence). Some are starting to enter the clinical market with shirts that do continuous heart monitoring, like Health Watch, an Israeli company that makes a 15-lead EKG heart monitor shirt. Used only for clinical purposes right now, the shirt can be worn continuously. It's monitored remotely via Wi-Fi or the cloud. This is great for hospitalized patients who need to have a heart monitor. Instead of wearing an uncomfortable external heart monitor that keeps you in your bed, patients can wear a monitoring shirt instead. You have a lot more freedom of movement, which makes a big difference when you're stuck in a hospital bed and already feel tied down by IVs and other tubes or devices.

Another company called Myontec makes athletic shorts that measure muscle activity in top-level athletes. If you can measure muscle activity in athletes, you can measure it in people with neurodegenerative diseases. That would be a major help in diagnosing and managing illnesses such as Parkinson's disease and muscular dystrophy.

Google has a project called Jacquard that weaves conductive metal threads into fabric, turning your clothing into a touchpad. You can't feel that the fabric is interwoven into your clothes. It's controlled by a chip the size of a jacket button. It functions like a smart phone screen. Imagine typing on your leg! The data is transmitted wirelessly. Eventually, Google is planning to converge Jacquard with technology that captures energy from motion, so your pants will charge while you walk.

To solve the problem of dead batteries—and my personal wearable devices seem to always be dead—get rid of the batteries. Tim Brownstone, the CEO of KYMIRA, started by making medical products that passively harvest wasted body energy and convert it into infrared energy. Now they're taking it a step further and developing textiles that actively harvest waste energy and convert it to electricity to power an array of devices, from consumer electronics through to biomedical sensors.

EPIDERMAL ELECTRONICS

We're about to see massive acceleration with what we can do with epidermal electronics—sensor-laden temporary tattoos that stick on the skin. They'll detect and transmit data via Bluetooth technology, including your temperature, blood pressure, and blood sugar level, and download it directly to your device with no human intervention. Epidermal electronics are non-invasive, painless, and unobtrusive. The tattoos could be worn for weeks. You'll be able to shower with them and do all your normal activities, including swimming.

The electronics company MC10 is working on epidermal conformal electronics, including ones that can tell you, via your smart phone, when to reapply sunscreen, your temperature, and your hydration levels. MC10 has partnered with L'Oréal cosmetics to do an in-depth study on human skin. L'Oréal will probably be launching the MC10 epidermal sensor in the near future, along with their makeup line, to tell consumers about their skin.

Sensor-laden temporary tattoos that are a supplement or replacement for skin is an area that's starting to get a lot of interest from scientists and entrepreneurs. At the University of Tokyo, researchers are working on creating electronic "skin" made of organic semiconductors and carbon nanotubes. Stanford is also working on skin for robots and robotic limbs that would be sensitive enough to feel a butterfly land on it. At UCSD, researchers are working with sensor ink! This would literally allow you to draw the sensors right on the skin. This means shape, mass, bulk, and permanence have all but disappeared. You're limited only by your imagination and creativity on how the ink gets applied.

Helping enable this technology, in 2015 the Department of Defense (DoD) poured $75 million into a consortium for developing flexible electronics, including 162 universities, nonprofits, and private companies. This will be supplemented with an additional $90 million from academic and private funding. Some of the organizations in the consortium are working on methods for manufacturing patches that cost significantly less than current methods. When a rapid, significant drop in price (to say, $1 per patch) arrives, this technology will suddenly be all around us. It will feel as if almost overnight, everyone has them.

We're also going to see massive acceleration in implanted sensors placed subcutaneously, just under the skin. You can't see them, you can't feel them, but they provide continuous, active body monitoring. Some people already wear tiny video cameras on their lapel for life logging and to track their health. TASER outfits police all over the world with mini cameras on their shirts to record everything when they're on duty. Why not extend that concept to patients?

The next step is to put the tracking device under the skin or in the blood vessels. You can't forget about it, and the battery isn't going to die. You can't lose it, and you don't have to worry about whether it's obtrusive or matches your outfit. Most importantly, a sensor inside a blood vessel provides continuous, accurate monitoring that the patient doesn't have to think about. Compliance goes way up when you don't have to worry about remembering to put a device on every morning, charge it, or download the readings.

Convergence Alert: Skin Sensors

Why stop with touch-enabled smart accessories and wearables? Flexible skin sensors can turn your body into a digital touch panel. An international team of researchers has come up with a way of attaching flexible touch controls directly onto the surface of our skin, as a very thin silicone overlay. Dubbed iSkin, the system does away with one of the fundamental problems of current wearables: finding practical places on the body to position technology that is both functional and ergonomic without being uncomfortable or too conspicuous (read: Google Glass). By using skin as the location for a near-invisible control overlay, iSkin's developers think they've discovered the perfect natural touch surface. The iSkin sensor is soft, flexible, and stretchy, not rigid like most wearables, and it can cover locations like the forearm comfortably. That gives a much larger input space than wearable sensors can provide.

Uses for iSkin sensors could include being a remote control mechanism for digital devices, such as answering incoming calls on a smart phone, controlling playback on a music player, or even typing and sending whole messages using printed QWERTY keyboard interfaces. Customized iSkin patches could be created and printed for personalized applications. Future versions might even source their power directly from the human wearer.

NewDealDesign, the company behind the original Fitbit design, says their implanted technology should hit the market around 2019 or 2020. Beyond basic biometric activity monitoring, it will also be able to open up keyless entry locks and even record who you've seen that day. It will be one of many unobtrusive, invisible life-logging devices and applications that we'll be seeing soon.

Implants are also making a huge difference for people with vision problems. Diseases such as age-related macular degeneration (AMD) gradually destroy the retina in the eye. Bionic eyes can help restore the lost vision. An electrical implant into the retina can be attached to a tiny camera worn on eyeglasses. The video feed is sent to the undamaged area of the retina. Interestingly, this means you can see with your eyes shut! Right now the system is a bit clunky, with a fairly large external power pack, but it's the start of a major change in treating vision-robbing conditions. Over time, the implants will become smaller; they may skip the eye altogether and go straight to the brain.

MONITORS AND MOTIVATION

One of the fun and motivating aspects of health monitoring devices is gamification. This is a technique that really works, and is great for engaging people in their own health. Let's say you need to get more exercise, and you decide to do it by walking on a treadmill. For most people, that's so boring that you probably won't stick with it. Add in a gamification component that you track with your activity monitor on your smart phone, and suddenly, it's fun. It's something you look forward to doing every day because now you're competing with other people, graphing your progress, comparing it to your friends and your family, and getting rewards for reaching a goal. Suddenly, you start to see the benefits of the exercise. We all like to win, compete, and solve. It's not competitive in the "win at all costs" sense; it's competitive in the sense of being enabling and motivating, with the sense of sharing, community, and feedback loops. When you know other people are watching, monitoring, and expecting you to compete, it's harder for you to skip your exercise!

CONSUMABLE TECH

We're also seeing sensor technology go from wearables to consumables. What does that mean? It means you swallow a pill with the sensor in it. We already have some FDA approved consumables (cameras you swallow) that are used to visualize the intestinal tract and replace the need for a colonoscopy. There's also a way to make sure a patient is taking medication as directed. Proteus Digital Health has created a first-to-market ingestible pill that contains a sensor the size of a grain of sand. In combination with a patch worn on the skin, the sensor pill can detect what drugs the patients take and when and then send that information to the patch. From there, it can be downloaded via Bluetooth to a smart phone. When the ingested sensor and pill hits the patient's stomach, the gastric juices create a chemical reaction that powers the sensor enough to transmit the data. Proteus has partnered with Otsuka Pharma; the first drug with an ingestible sensor (Abilify) has now been approved by the FDA.

Scientists are creating an ingestible sensor that measures heart rate and respiratory rate, using technology created at MIT. It measures these two vital signs by recording the sounds of the heart beating and air moving in and out of the lungs, all from the GI track. It's about the size of a vitamin and contains a tiny microphone and electronics inside a silicone capsule, and it sends radio signals to an external receiver. Among the many uses this technology will have, it will be able to replace the uncomfortable heart monitor harness patients need to wear for a week to diagnose some cardiac diseases. In this particular case, we'll see compliance go way up, as swallowing a pill is much easier than wearing a harness for 24 hours a day for a week!

The Google life sciences group is developing a smart pill that could scan for cancer. The pill is packed with tiny magnetic particles that seek out malignant cells in the bloodstream and report the findings to a wearable device. The same smart pill technology could be used to detect early risk factors for other problems, such as kidney disease.

Propeller Health recently received FDA clearance for a device that attaches to an asthma inhaler and uses GPS to track the inhaler's use in real time. The device helps improve medication adherence, which is important for asthma treatment. The information from the device is also voluntarily anonymized and aggregated for research in improving asthma management.

BRAIN TECH

Brain-computer interfaces (BCI) use sensors that can read and record brain waves with noninvasive, inconspicuous sensors that fit on the head like caps or visors. Ariel Garten, CEO of InteraXon, has created a comfortable, lightweight headband called Muse. The device tracks your brainwaves as you meditate and gives you real time feedback on your smart phone. Emotiv is working on a headset that gives EEG readings, including real time data and analysis on what the user is interested in—stress levels, attentiveness, excitement levels, and more—for only a couple hundred dollars. Previously, this technology was so expensive and difficult it wasn't accessible to the average consumer in the US. Now, it's the cost of a couple nice dinners.

Neuroscientist David Eagleman has invented a noninvasive medical device that allows deaf or hearing impaired patients to hear, no matter why they can't hear naturally. This is a game changer for so many people. The medical device is a vest that works with a smart phone or tablet. What happens is that one person talks, the smart phone or tablet records it, and turns the sentence into a series of signals that are then transmitted to the patient's vest. The vest in turn transmits the signals to the brain by using a series of vibrations, completely bypassing the ears. The brain does need training and practice to receive and understand these new sensory signals. David has an awesome TED Talk that explains the technology in greater detail.

Adam Gazzaley's lab at the University of California, San Francisco is a cognitive neuroscience research lab focused on studying the neural mechanisms of memory, attention, and perception. For some of the studies, participants wear a sensor cap that fits comfortably over the head like a ski cap. This is a huge step forward from measuring brain activity with EEG wires attached to the scalp at one end and a large recording device at the other. Gazzaley is also working on video games to treat conditions like post-traumatic stress disorder, traumatic brain injury, attention deficit hyperactivity disorder, autism, and Alzheimer's disease.

DARPA's ElectRx project is a $78.9 million research program to develop minimally invasive neurotechnologies—microscopic chips that modulate peripheral nerve circuits. According to DARPA, the goal is to understand the structure and function of specific neural circuits and their role in health and disease. The chips would allow doctors to diagnose, monitor, and treat illness and injury. By continually assessing conditions and providing stimulus patterns that help maintain healthy organs, the chips would help patients get well and stay healthy using their body's own systems. The main target would be brain and mental health disorders such as epilepsy, traumatic brain injury, post-traumatic stress disorder, and depression.

The ElectRx device is like a miniaturized pacemaker. It would send electrical impulses to monitor the body's organs and heal damaged body parts without using drugs. Because it's so tiny, it can be placed on nerve endings exactly in the right area.

At some point in the future, BCIs will be routinely used to control prosthetic and robotic limbs. Scientists have already been successful with this technology, with a couple of famous cases from 2012, including one where a paralyzed women was able to drink out of a straw by herself for the first time since she became completely paralyzed, using a BCI to control a robotic arm.

You can think about moving your prosthetic arm to pick something up, and it will. BCIs will even super-enable people. You could wear an exoskeleton robotic limb or entire suit, controlled by a BCI. Not only is this a huge help for people who have muscle wasting from old age, degenerative disease, severe illness, or medication but it also turns us all into superheroes with super strength! Maybe you'll put on a robotic exoskeleton suit to make yourself superhuman, or just use a robotic exoskeleton arm. (I'll talk a lot more about robots in chapter 6.) Multiple companies are working in this space, already supplying exoskeletons for physical therapy, paraplegics, military use, and even in factories to help workers be stronger.

A brain-computer interface has recently been used experimentally to restore the ability to walk to a patient with a spinal cord injury. The system uses an EEG cap to read brainwaves and then transmit them as electrical signals that travel down to electrodes placed around the patient's knees; the signals generate movement and enable basic walking. So far, the system has only been tested on one patient, but the results are very promising. Researchers are already looking ahead to using a brain implant instead of the EEG cap for even greater control and possibly even restoring sensation to the paralyzed legs.

And taking it a step further, maybe your BCI will control the personal robot that functions as your in home healthcare aide, butler, and maybe even cook. Wouldn't that be great for people who are paralyzed, visually handicapped, or can no longer speak? Or for those of us too lazy to get off the couch? You'd be able to just think about having another slice of pizza, and your home robot would bring it to you.

The University of Southern California is in the early stages of testing implanted brain chips to enhance, supplement, or replace long term memory retention, especially helpful with Traumatic Brain Injury (where your head has had an impact that causes damage), Alzheimer's, or even just natural aging. They have already seen success with their BCI chip in rats and monkeys and are now testing it on epilepsy patients.

TMI: TOO MUCH INFORMATION?

In the future, you'll be able to collect vast amounts of information about your health. What will you do with it? David Ewing Duncan is called the "Experimental Man." He has done more tests on himself than almost anyone else in the world! He has over ten terabytes of data on himself drawn from biometric tracking, MRIs, CT scans, blood tests, DNA sequencing, and more. How do you sift through this information and use it to see not only the big picture but also the individual problems?

Larry Smarr is the founding director of the California Institute for Telecommunications and Information Technology at UCSD, and he is called the "Patient of the Future" because he ended up self-diagnosing his ulcerative colitis. Smarr tracks more than 150 parameters about his health. Some he measures continuously in real time with a wireless gadget on his belt. He logs his weight daily. He tests his blood and the bacteria in his intestines once a month. Is this excessive? Smarr doesn't think so. He says he sees his body like a car: "We know exactly how much gas we have, the engine temperature, how fast we are going. What I'm doing is creating a dashboard for my body."

Platforms to manage all that tracked information are starting to gain popularity and market share. Right now, there's a lot of competition among the big tech players to be the platform everyone uses and builds on to be compatible. One example is Qualcomm Life's 2net platform, a cloud-based system that captures, transmits, and aggregates biometric data from medical devices and sensors. It allows healthcare providers, caregivers, and patients to all have access to the data. Data sent to the 2net platform can be also synced with Qualcomm Life's HealthyCircles Care Coordination Platform. This software helps with care coordination. When sensors from someone with heart failure, for instance, show a jump in blood pressure, the system triggers an alert for a healthcare provider and any caregivers to check in on the patient.

If you have diabetes, continuous glucose monitoring means you'll never have a hypoglycemic episode. If your blood sugar starts to get into the danger zone, you'll get a warning text from your smart phone, an immediate release of insulin from an artificial pancreas, or a robot delivering you a cookie without you having to do a thing. You'll also have a much better understanding of how to avoid a low or how much insulin you need if you're about to exercise because all your food intake and physical activity will be tracked 24/7. By analyzing that data using artificial intelligence, you'll be able to come up with insights about your blood sugar that you can't get with today's monitoring. Similarly, if you're doing chemotherapy, continuous monitoring can help you catch dangerous side effects before they get serious. Likewise, if you need to take a drug that has a narrow line between the optimal dose and an overdose.

We'll soon be able to monitor pretty much every body activity continuously, effortlessly, and passively. Your internal sensors will silently gather information and store it in the cloud where it will be monitored by artificial intelligence.

If a problem arises, you'll get an alert. Years could go by without your ever getting one, but you won't notice or care because you won't even think about it. It's not even file and forget because when the sensors are inside the body and woven into our clothes, we won't even need to remember to forget it. It's going to be as natural as breathing.

It's possible that at first, all this data may lead to false alarms. We'll need some time to gather enough data from enough people to analyze it accurately and avoid misdiagnoses. But when you add together biometric sensor data with individual genome sequencing and medical history, we're going to be able to come up with remarkable insights based on pattern analysis. With the Internet of Things (IoT) collecting data from Smart Cities, weather, environment, where you go, and what you do, we can start to layer that over medical data to come up with an even more complete picture.

We have no idea what that picture will look like or what it will tells us because no one has ever done continuous health monitoring on a really large group of people before. We know that large studies are valuable because we can look back to the original Framingham Heart Study, which began in 1948 and looked at the factors affecting the heart health of about 5,200 adults in Framingham, Massachusetts. That study and the follow-ups to it have provided a lot of insight just from a relatively small data set.

Imagine what will happen when we can study the data from millions or even billions of people. That time is approaching fast. We've already seen one good example of how the power of aggregation can reveal important information. Practice Fusion is an electronic medical records company that is both patient-facing and doctor-facing. It works across all devices and is always free. It's the largest cloud-based EMR company in the US. The beauty of managing the records 55 million patients is that the anonymized information can reveal unexpected links. Recently, an analysis of Practice Fusion records showed that taking proton pump inhibitor medications, such as omeprazole (Prilosec) and lansoprazole (Prevacid), for acid reflux may be associated with an increase in the risk of having a heart attack. If you're taking these drugs, this information could save your life.

A lot of what I've been saying about monitors and devices is valuable for someone with a chronic condition, but it's also very valuable for preventive medicine. We're moving in that direction more and more because right now 80 percent of our healthcare dollars are spent on only 20 percent of the population. Most of the money goes for chronic disease care and terminal care. More of it needs to go to preventive care. If you can prevent a disease or catch it early on, it's much more treatable and even curable. That works for cancer, but think of it in terms of every single disease out there. Imagine a world where we strive for not just Stage 1 but Stage 0 medicine.

Chapter Two

PORTABLE MEDICAL DEVICES AND POINT OF CARE DIAGNOSTICS

RIGHT NOW, IF YOU NEED AN ULTRASOUND, A COLONOS-copy, a urine analysis, a blood draw, or any other medical test, you usually have to go a lab, a testing center, or a doctor's office to have it done. A lot of other tests, such as an eye exam or a breast exam, are done by clinicians, usually in the office setting.

A point of care diagnostic tool, or what I like to call a "diagnostic on demand," is a medical testing device that comes to the patient, as opposed to the patient going to the diagnostic tool. A point of care diagnostic can be administered by someone who's not a physician or by the patient.

Of course, as an individual, you can't just take an ophthalmologist's tools and look in someone's eyes and be able to diagnose any-thing. But with the advent of the client care diagnostic, you don't need an ophthalmologist. A technician trained in using the device will do the testing, but the results will be sent to the cloud. A physician, probably assisted by AI, will then read them. That physician could be anywhere in the world.

For eye exams, we already have a company called EyeNetra, cofounded by Ramesh Raskar of MIT Media Lab. EyeNetra makes a device that attaches to a smart phone to do refractions, the eye exam that determines if you need glasses and what the prescription should be. Right now, you need to go to an optometrist or ophthalmologist for a refraction. The EyeNetra device is small, lightweight, and portable, and it can be use by anyone, anywhere. In other words, basic vision correction could be provided easily to the 2.4 billion people in the developing world who can't access it right now. This is a great example of how exponential technology can leapfrog medical care ahead. Just as smart phones are being widely used in underdeveloped parts of the world that never had landlines, this technology leaps over the lack of traditional eye care to bring fast, easy, inexpensive care to people who never had any kind of access to it before.

Those tricorders they use in the *Star Trek* movies to diagnose and treat people aren't just science fiction anymore. Small, accu-rate devices that painlessly measure or diagnose body functions are here already. The Scanadu Scout device, about the size of a dental floss package, contains sensors that take your tempera-ture, blood pressure, and pulse just by holding it briefly to your forehead. The Scanadu Scanaflo urine analysis kit will run a dozen standard urine tests using a simple dipstick system similar to a pregnancy kit, and it will cost about the same. The results are sent to the cloud to be analyzed by holding your smart phone camera over the dipstick result's barcode. So, instead of going to a lab or the doctor's office for basic urinalysis, you can do it quickly and conveniently at home or at work!

Tympani, made by Caring Things, is an infrared thermometer that connects to your smart phone's audio jack to take your ear temperature. Because it works through your smart phone, Tympani doesn't need a separate battery or display, so it's very compact. It takes only a few seconds to use the device to measure the temperature inside the ear—great for getting the temperature of a crying baby.

Eko, cofounded by Connor Landgraf, makes an FDA cleared device that fits into a standard stethoscope and lets a healthcare provider make a digital recording of your heart sounds using Bluetooth technology. The recording can then be sent to a specialist for further review. That gets a quick diagnosis without a second trip to a specialist.

The tricorder concept—a small, portable, wireless device that monitors and diagnoses health condition—is the idea behind the Qualcomm Tricorder XPRIZE. This incentivized prize will award $10 million to the team that comes up with the most innovative design for a portable device that measures five vital signs and diagnosis 15 diseases in the field without the presence of a doctor. I love the concept of incentivized prizes. They're designed to catalyze breakthroughs by awarding money to a winning idea or team that solves a defined problem. They encourage people to work on a specific problem within a competitive environment.

Biogen, a biotech world leader in Neuro, has been able to make a PET Scan machine that only costs $200! Until now, a PET Scan machine cost in the millions of dollars and was huge. Only hospitals could afford it, and the developing word definitely could not afford it. With Biogen's incredibly innovative technology, a PET Scan costs pennies to administer using their new small, portable machine.

THE FUTURE ON YOUR SMART PHONE

Many of the technologies that will transform medicine already exist and are being used now. Several companies, for instance, offer heart monitors that fit onto a smart phone. The AliveCor EKG monitor is just one of the many EKG monitors on the market. Others can track heart rhythm symptoms, activities, and diet to help determine the cause of heart rhythm disturbances. To record the EKG, a small, mobile device is attached to the phone or tablet. All you have to do is place two fingers of each hand on the electrodes. The EKG is recorded through the microphone and sent to the cloud, where it's read by a cardiologist (I can imagine eventually an AI cardiologist). The app stores the information. The cost? Under $100. If you have a heart condition, this is great way to monitor it. And if you live somewhere far away from a cardiologist or want to go on a vacation after having a heart episode, you can still access great care. You can even get a version of the AliveCor smart phone app for your dog. I'd advise against getting your pet an actual smart phone, however.

Imagine going to your doctor and getting prescribed an app instead of a medication. Beyond the tracking and life-logging apps that can replace needing to keep a journal of exercise, food intake, vital signs, and more, the world of diagnostics on demand is expanding rapidly. Your doctor might prescribe the Withings blood pressure monitor or the Masimo IP Oximeter to measure oxygen levels in the blood. Many other devices that work in combination with apps that store information in the cloud are now available. Quite a few of the apps are approved by the FDA, which says that "medical devices that are mobile apps meet the definition of a medical device and are an accessory to a regulated medical device or transform a mobile platform into a regulated medical device." If an app has been approved by the FDA, that means a physician can potentially prescribe it for patient or allow it to be integrated into the patient's EMR (electronic medical record).

CellScope is one of the Rock Health portfolio companies. Their mission is to create the world's first smart phone-enabled digital tool kit, designed to be accessible with easy-to-use tools that can capture diagnostic-quality data. CellScope uses a small device to turn a smart phone into a tool for an ear examination. The device captures video of the eardrum and sends it to a doctor for a diagnosis. This is particularly helpful for handling kids who get frequent earaches and are often prescribed unnecessary antibiotics for them. The video can be sent to one of CellScope's 24/7 on call physicians who will respond with a virtual consultation within two hours, or it can be sent directly to your regular physician. Companies like CellScope are empowering patients by shifting the power of administering the exam, owning the exam data, and choosing the providers. Devices like this improve access to care and increase the likelihood of catching medical issues faster and at an earlier stage. The potential to dramatically improve outcomes is very large.

Incubating Innovation

Business incubators are organizations that help start-up and early stage companies develop and grow. Rock Health, founded by Halle Tecco, is a nonprofit incubator in San Francisco that helps facilitate and encourage innovation in digital health that benefits patients. Other incubators that have really helped empowered entrepreneurs to make patient lives better include Saeed Amadi's Plug and Play, a co-working space and incubator. Amadi has partnered with multiple large healthcare organizations to help get innovations distributed quickly. Y-Combinator and Dave McClure's 500 Startups are other healthcare innovation incubators. I like to keep an eye on what these and other incubators are doing. It's a great way to see what new ideas are being worked on in healthcare.

Some devices will empower you to make an immediate decision on your own. Other devices will change the patient experience by letting you avoid a trip to the doctor to diagnose the problem. Some devices will help you manage a chronic condition better by tracking it and analyzing it.

For example, the uBox is a smart pill box that reminds you to take your medication. It's portable and flexible. For people who have trouble taking their meds on schedule, it has a component that can alert family members, caretakers, or healthcare workers if you have or haven't taken them. This is a big help for helping people comply with their medication regimen, especially if it's a complicated one or if they have cognitive problems. Or even if they're like me because I don't always remember when to take my medication or if I have already taken it that day. Sometimes I'm holding a pill bottle and can't remember if I took the medication in the past five minutes!

The XploR mobility cane for people with visual handicaps uses smart phone technology to recognize familiar faces from up to ten meters away. It has a GPS function to aid navigation. One of the hardest things to do when you have a chronic disease or limiting health problem is live a normal life. The XplorR cane is a great example of how accelerating technology can be very enabling and restore independence.

Twenty years ago, you needed a mainframe computer to get the same computing power that's now in your smart phone. Today, you can diagnose a serious medical problem by attaching a dongle to the audio jack of your phone—perhaps a device that listens to your heart or one that contains reagents in a disposable plastic cassette for doing a urine analysis. The whole thing is powered by your smart phone. And in many cases, your phone is smart enough to do all the analysis on the spot without having to send the data to the cloud. The AI component is being programmed into a lot of apps. You can still send the info to the cloud for storage, aggregation, and compilation with other people's data and a more detailed trend analysis, or you can simply transmit the results directly to your doctor.

Qloudlab, a Swiss company, is working on a way for people on antiplatelet therapy such as the drug warfarin (Coumadin) to monitor their blood levels of the drug at home with a single-use film attached to the screen on a smart phone. If blood levels of the drug aren't within safe limits, the information can be sent to the doctor and treatment to avoid a major bleeding episode can be started. Right now, patients on these drugs use a coagulation meter with test trips to check their levels. As with glucose meters, these devices are clunky, inconvenient, and require a painful finger stick. Plus, they're expensive. Each test strip costs about $12. Bleeding caused by these drugs is a common cause of emergency room visits and leads to about 33,000 thousand hospitalizations a year. Studies show that people who test their drug levels often reduce their risk of a major bleeding event by about a third. What if they could use the Qloudlab device, which requires a very tiny amount of blood, instead? Compliance would go up, and bleeding episodes would go down. The savings would be huge because the average hospitalization for a bleeding event lasts six days.

Mental health can be tracked and diagnosed too. The Ginger. io app uses sensor data collected through your smart phone and self-reported information that can identify if you need mental healthcare. We're all on a mental health spectrum, so everyone could benefit from that sort of tracking. We all get situational depression, and we all go through different moods. The app can alert you that maybe you need to talk to somebody, or in my case…go on a shopping spree. If you have a diagnosed mental health disorder, the app can help note if things are getting better or worse and if maybe you should talk to your healthcare provider about starting or changing medication or dosage. Ginger.io is another example of how passive sensors that simply collect information about what you do all day can be used to monitor a condition and send alerts if needed. The sensors give you insight and actionable data. It's a great example of personalized medicine; it's passive and unobtrusive, yet it can tell you things about yourself and alert you to a potential problem early on.

Is there really a need for this sort of app? Yes. One in every 100 people between the ages of 14 and 27 are at high risk for psychosis, for example. An automated computer program developed through IBM Watson identified, with 100 percent accuracy, at-risk patients who went on to develop psychosis (losing touch with reality) from those who did not. Long before a doctor might detect it in an interview, the at-risk individuals were stringing their words together in a subtlety abnormal way. Watson uses algorithms to identify these abnormalities very precisely and objectively.

A computer program called Ellie is designed to diagnose post-traumatic stress disorder (PTSD) and depression using AI. While the program asks you standard diagnostic questions about your family, how you're feeling, and so on, it's also reading your facial expressions and analyzing your tone of voice. In tests, Ellie diagnoses people with PTSD correctly more often than trained therapists. One reason it's more accurate is that it doesn't give patients feedback, something therapists may unconsciously do even though they're trained not to. When you take the therapist's response out of the equation, patients report that it's actually easier to talk openly to a nonjudgmental machine.

We now already have a lot of individual devices and apps to actively monitor individual health factors like your blood pressure or blood sugar. This is the interim step. Some of the devices are meant for sophisticated users who can afford them, but a lot of current and near future devices will be used by all patients who have a specific health issues. For example, your doctor will prescribe, and your insurance company will cover, a lab-on-a-chip using your smart phone so that you can test your anticoagulant drug levels every day.

Healthcare workers will be able to test for infectious diseases on the spot quickly and inexpensively using pieces of paper. GE has partnered with the University of Washington to create a paper based microfluidic exam to test for infectious disease in the field. Results are available on the spot in under an hour. But you don't have to be a big company like GE or a major research center like University of Washington to be innovative. Olivia Hallisey, a 16-year-old from the US, won the 2015 Google Science Fair with her project to develop a fast, cheap, and stable test for the Ebola virus. It gives easy-to-read results in less than 30 minutes, potentially before someone is even showing symptoms.

What is amazing about technology is it makes things seamless. Where future technology will take us is to continuous, accurate, passive monitoring. We won't have to actively do anything. It will alert us, our physician, or our caretaker when problems start.

We'll know about the onset of some small symptom before we even feel it a lot of the time. These technologies are showing how that's going to eventually be the norm and the standard of care. When passive monitoring becomes the norm, we're suddenly going to have massive amounts of data that may completely rearrange the way we look at things. What is high blood pressure really? Is it the number you have in the doctor's office when you might be feeling sick and not all that relaxed, or is it the number when you're at home reading a book? Passive monitoring will give you the bigger baseline picture and let you see how and when fluctuations happen.

Sensor technologies and point of care diagnostics and monitoring devices are applicable across every single disease, condition, and syndrome, even for prevention. Eventually, we'll have technology that will track it so that you will be able to manage your own body. That gives you massive empowerment. If you have access to all this data, and you're measuring things constantly but passively without having to put any thought into, suddenly you're empowered to take action on your own health. It's like having a smoke detector for your whole body. You don't notice it's there unless it goes off. And when it does, you know there's a problem.

At-home monitoring based on blood draws is always a problem. Those finger pricks hurt, which keeps people from doing them often enough. For a bigger blood draw, you might have to get yourself to a lab regularly, which can be a real problem in terms of time, transportation, and cost. Plus, vein punctures hurt! Sometimes the phlebotomists don't hit the vein on the first try...or worse, they hit an artery instead! I once had a blood draw that sent a small but powerful geyser of blood shooting out of my arm a foot into the air! Any sort of skin puncture for blood opens you up to the risk of infection, especially if you're immunocompromised. Needle-free blood draws are on the near horizon, using capillary action to pull blood from a finger tip. The HemoLink device for self-collecting blood samples works exactly that way. You collect your own tiny blood sample at home and mail it off to the lab for analysis. Of course, in the future, you'll do the analysis yourself using microfluidics and a lab-on-a-chip attached to your smart phone or at a lab in your local pharmacy (see chapter 8).

The technology is accelerating ever faster. Within a short few years, we're going to see more devices and more point of care diagnostics in our hands because massive amounts of money and resources are being put into them. For example, Walgreens has partnered with Qualcomm Life, a division of Qualcomm with expertise in sensors and computing systems. The collaboration will collect all your personal health data from all sources into one cloud-based platform. That really enables medical devices and apps from end to end. Wouldn't it be great to actually have all your healthcare data, from whatever source, in one place, instantly accessible by you and your healthcare team? Beyond that, the massive aggregation of millions of records could provide valuable insights into all sorts of healthcare issues. Unexpected drug side effects, for instance, might not be noted individually, but if hundreds of people report them and the information is aggregated, then the problem pops out of the data.

Walgreens is a massive pharmacy chain. If you take health-care records to the next level, Walgreens might offer a rewards program for buying a fitness wearable. An incentivized program would encourage you to report your results and achieve your health and wellness goals.

The medical lab, whether it's in a hospital or stand-alone, will eventually become a thing of the past. Technology will redefine where lab tests occur and how they're done. Microfluidics and needle-free blood draws will make the process much faster and easier and let most testing take place at home, by the bedside, in the medical office, or at a pharmacy.

A big point of all this technology is that it will enable physicians to do more of what they're better at: treating their patients. No longer will they have to wait days for test results or track down results that get lost. They'll be able to help patients manage their health better and avoid emergency room visits and hospitalizations.

I'm more than willing to be a guinea pig for new devices. When the Quantified Self movement started in 2007, I was on board as soon as I heard about it. In fact, I worked for a company (HealthTap) that was the first annual sponsor for the Quantified Self back in 2010. Personally, I am totally looking forward to sensing up my body. I want to have subcutaneous sensors, sensors on my brain, and sensors inside my blood vessels. I want sensors that will measure and continuously monitor my heart and my brain waves, all my blood levels, liver enzyme levels, sleep stages, everything. Sensors are getting smaller and smaller and cheaper and cheaper. When we talk about putting a nanotechnology sensor inside a blood vessel, that is tiny. You could literally have hundreds throughout your body and not even know it.

Diabetes Revolution

Amazing advances in treating type 2 diabetes are a great way to show how technologies are converging to impact a disease—in this case, the fastest growing condition in the US. About 29 million Americans live with this disease. And for every person who knows he or she has diabetes, there's another who's on the verge or already has it and doesn't know it. Monitoring your blood sugar using a glucometer and test strips is crucial for managing the disease. Today, people with diabetes need to prick their fingers several times a day to get blood sugar readings. The finger stick is painful, inconvenient, and expensive; the test strips cost about a dollar apiece. Some people with diabetes should be tested four times a day, but health insurers often will pay only for twice a day. That leaves the patient the choice of not testing often enough or paying for test strips out of pocket. The finger stick approach is a big reason why a lot of people with this disease have trouble being compliant with their lifestyle changes and medication.

Technology is changing this as wearable, needle-free blood sugar monitoring becomes a reality through contact lenses and other innovative ideas, such as temporary tattoos with sensors and low-power lasers that detect blood sugar levels through the skin. Patients will use their smart phones to track their readings and get advice about their diet and exercise. The information will be sent to the cloud where users and their healthcare team can download the information. When the wearable blood sugar monitor becomes available, it will be amazing for people with diabetes.

Diabetes Revolution

They'll be much more willing and able to manage their disease, which in turn will help avoid the devastating and very expensive complications of diabetes, such as amputation and blindness.

The wearable meters will calculate insulin doses and warn people of potential low blood sugar episodes in time to stop them. The cost impact will be huge. Problems from insulin are a major cause of drug-related emergency room visits, and simply cutting back on these would save a lot of money. Diabetes-related complications cost billions every year.

A coming nanotech solution for people with diabetes who need to inject insulin is a smart patch lined with painless microneedles full of insulin. The needles are the size of an eyelash; so fine that when they inject the insulin, you don't feel it. When this becomes available, life for many people with diabetes will become much more comfortable.

Chapter Three

INFORMATION TECHNOLOGY AND COMPUTER SYSTEMS

TODAY, WE HAVE ACCESS TO AN AMAZING LEVEL OF INFOR-mation about our health. That's really changed things, especially from the patient's perspective. We can now access information from anywhere, not just in front of a computer terminal. Where once you needed to visit a research library or medical school library to find detailed medical information and journal articles, now you can do it all online. When I first got sick as a teenager, most of my information that didn't come from my doctors came from the massively thick and confusing *Physician's Desk Reference.* It almost outweighed me, I didn't understand most of the medicalese, and there certainly wasn't a control F function to find all the cross references. The old-fashioned approach of looking something up in a thick, incomprehensible medical reference book such as the PDR or *Merck Manual* has been replaced by searching Dr. Google.

All that information on the Internet can be incredibly valuable, but it can also be dangerous. According to Pew Research Center surveys, 72 percent of Internet users say they looked online for health information within the past year, and one in 20 Google searches are related to health. The surveys also say 35 percent of Americans Google their symptoms. But when it comes to health, the Internet is unfortunately very full of too much misleading, biased, outdated, and incorrect information.

This is a big issue because too many people can't distinguish between the good information and the dangerously bad information. For accurate, actionable information, you always need to use common sense and start with only the most reputable health sites. Wikipedia isn't always reliable. While Wikipedia articles usually cite references, those references aren't always 100 percent reliable. Product sites are consumer-oriented and not always written by actual healthcare professionals. Websites sponsored by government agencies, such as the many branches of the NIH (National Institutes of Health), are generally highly reliable. So are general sites run by major health institutions like the Mayo Clinic, the Cleveland Clinic, and disease-specific organizations like the American Diabetes Association. Hospital systems often provide accurate information. Drug company web sites are often good sources of information about their products and the overall disease, but it's important to remember that they're going to be biased in favor of what they sell. Google is now rolling out health information for about 900 diseases and conditions, and that's just to start. The easy-to-read search results include fast, at-a-glance information and easy-to-download pages to print out and take with you to your doctor. The results will be validated and accurate; gone will be the fire hose of information that searches now provide.

In 1970, medicine recognized about ten specialties. Today, we have over 170. No doctor, no matter how dedicated, can keep up with every new development in his or her subspecialty, much less know what's going on in other areas. Even doctors use Google.

THE POWER OF ELECTRONIC MEDICAL RECORDS

In the US, more than 40 percent of patients would switch doctors if that meant they could access their Electronic Medical Records (EMR). In fact, many patients already do have access to their records because EMRs are the norm now. Paper records and medical charts are being replaced with electronic ones at every level of healthcare. Patient portals, for example, let you access your medical records and access things like your lab results directly at any time, without having to call the medical office and request them. Your patient information is digitized in HIPAA compliant ways, so it can be sent instantly from one place to another. The same is true of your prescription records from your pharmacy.

At times, this might be too much information. Looking at your lab results without someone who can interpret them for you could send you into a panic. You might misinterpret something, become overly concerned about a small change in a number, or just not understand all those abbreviations and acronyms. That's where AI will come in. You'll be able to access the help you need from an AI "doctor" to understand your results and answer your questions—even when you're feeling anxious at 2 a.m.

If you want to take a deeper dive into the research, AI will help. When AI becomes much more accessible, it can sort out the research for you and then present you with the data and answer your questions about it. We're not there yet, but it's coming.

Over the next five to seven years, some three billion more people will be coming on line. Their medical information will become part of a massive data set. Vast amounts of your personal health data can be stored in the cloud and aggregated with the data from millions, eventually billions, of other people. The convergence of all that data will give us huge insights into health and disease. When you have that much information and you can analyze it with the computing power and methods we now have, that's very conducive to scientific discovery. We don't know yet what we'll find, but it's a fairly good bet that we'll be able to see patterns and influences that can show up only when large amounts of data are studied. We'll learn a huge amount about predictive analytics and early-onset diagnostics that will help with the leap toward Stage 0 medicine.

A big issue right now with EMRs is that there's no universal platform. The various EMR systems aren't compatible—they can't talk to each other. The Department of Health and Human Services has identified compatibility as a serious problem and is offering some incentive prizes for fixes. Even hospitals that use the same system can't always communicate well because each hospital tweaks the software for its own purposes.

What's a problem for the healthcare system is an opportunity for entrepreneurs. For example, a company called PicnicHealth will get all your records across multiple hospital systems for you then aggregate them into one, easy-to-read portal for you and any healthcare provider you designate. Noga Leviner founded PicnicHealth after having a hard time tracking all her data for her autoimmune disease. It's awesome when patients work on solving huge medical problems!

Companies like Google Fit and Apple's HealthKit and ResearchKit are putting massive amounts of resources into developing ways to aggregate data across wearable tech. If that information can be integrated into EMRs, then the data becomes even more interesting. Eventually, when we have clinically relevant sensors inside our bodies, that data can also be aggregated. Add in environmental data and even more interesting things can be discovered. If you have rheumatoid arthritis, for instance, AI could not only predict your health but help you manage your day-to-day life so that you can be as healthy as possible. AI will look at all your data from every source and tell you that you have the least pain on days when the temperature is 75 degrees, that so far you're not showing a signs of side effects from that new drug you started taking six weeks ago, but a particular food you ate on Thursday is now making your arthritis flare.

We're not there yet. In fact, we face a lot of problems trying to get there because not all the data aggregates well. The data components need to come from a lot of different places, but a lot of it is siloed—the different sources don't talk to each other yet. The artificial intelligence component isn't there yet. But the goal is there, and the problems will be solved.

UNLEASHING THE POWER OF THE GENOME

Both Google and Amazon are now storing genomic data in the cloud for you for the low, low fee of $25 a year. But wait, there's more! Your genetic analysis takes only 26 hours to do! When enough people participate in having their genes sequenced and shared, we'll have a massive database of genomic data that can then be analyzed to detect patterns and anomalies. The Human Genome Project and commercial companies such as 23andMe are already continually adding to our understanding of our genes. Having your genome sequenced today is far less expensive than it once was and getting cheaper and more accessible all the time. In fact, the price is dropping faster than Moore's law would predict. Once the price drops to just a few dollars, which it will probably do within a decade, we will be able to afford to sequence the world. When we start talking about the penny genome, everyone could get genetically sequenced at birth.

That will create a massive data set for doing basic pattern analysis. It will also let us predict what's going to happen to you in your life. We've already identified specific genes that are directly tied into certain diseases. If you carry some variants of the BRCA gene, for example, you're more likely to get breast cancer. If you do get cancer, genomic sequencing of your tumor cells will reveal the specific mutations they contain and tell your doctors which drug to use to target the mutation. This approach is already being used for a limited set of conditions; it will be routine for every cancer and other serious conditions not too long from now.

Genetic testing to determine your individual response to medications will also become routine. We already have a genetic test to determine your response to the blood-thinning drugs warfarin and heparin. Instead of adjusting the dose using trial and error, and hoping to avoid a bleeding event in the meantime, the genetic test will tell you if these drugs are safe for you and in what doses.

Genetic testing will also help us predict your lifetime risk of specific diseases. A gene test at 30 could predict your exact risk of developing type 2 diabetes by 45. That gives you plenty of time to start taking preventive measures. I'm at a 40 percent risk for diabetes when I get older. Because diabetes can largely be prevented by having a healthy lifestyle, I workout almost every single day to mitigate my risk (as well as other reasons), and I've done this for over 30 years, no matter how I'm feeling (with the exception of when I'm hospitalized). Other diseases might be prevented or delayed by beginning drug treatment before symptoms start to appear.

You can take genetic prediction a step further and do genetic alterations to remove the risk of a disease. If genomic testing shows you have a genetic predisposition for something, maybe we go in there and just fix that gene. It might be done at birth or even in the womb. Or it might be done at conception in a test tube to eliminate a deleterious gene that could cause a serious condition, such as spina bifida or sickle cell anemia.

Does this raise ethical questions? Yes, definitely. Designer babies are a huge ethical issue. Remember the movie *Gattaca?* Is it morally acceptable to tweak a fetus? While genetic modification brings lots of ethical issues, it also brings lot of opportunities. The first full sequencing of the human genome was done only in 2001, and it cost $2.6 billion. It took us years to get where we are now with sequencing, but the pace is accelerating dramatically. Inexpensive genome sequencing is such a massive breakthrough that it will lead to all sorts of things, including some that will be completely unexpected and unpredictable.

VIRTUAL REALITY GETS REAL

Virtual reality (VR) isn't weird futuristic stuff anymore. It's here. VR is being used in surgery and in medical education. It's also great for patients. A patient might wear a VR headset at home, and a doctor could do a virtual house call using video from a computer. It would be an immersive experience that would be effective and much more convenient for everyone.

A lot of money is poured into virtual reality. Oculus Rift, a maker of VR headsets, was acquired by Facebook in 2014 for $2 billion. Google offers Cardboard, an inexpensive, do-it-yourself cardboard mount that turns your smart phone into a virtual reality platform for under $30. When VR is this cheap and this much fun, inventive new ideas are certain to come from it.

This technology has dramatically decreased in price, and it is massively increasing in how good it is. The immersive experience of Second Life, created by Philip Rosedale, creates a virtual world. High Fidelity is Rosedale's next generation virtual reality online environment. Soon patients will put on VR headsets not just to see a healthcare provider but to live their lives. Chronic disease patients or people with severe disabilities will be able to leave their homes in virtual reality. This gives them a massive level of independence. They can now go to work and be in an actual meeting in a room where everyone is virtual and everyone is wearing VR headsets.

VR can provide a new level of independence and the ability to maximize living your life, which is a game changer for patients. If leaving the house is hard for you, you can still socialize using VR. An elderly grandmother can use VR to see her great-grandchildren.

Imagine adding haptic technology to VR technology. Haptic technology provides tactile feedback to create a sense of touch through vibrations and other motions. Let's say you wear your VR headset and put on haptic gloves. When you reach out and touch something in the virtual world, you get the same sensory input you would get in the real world.

A very practical use for VR is pain control. The distraction provided by the VR scenario can help reduce the use of powerful and potentially addicting drugs. The value of this has been shown with burn patients, who suffer very severe pain on a regular basis when their bandages are changed. An immersive VR game called Snow World has been tested and shown to help reduce pain for young burn patients by transporting them to a polar world of penguins, icebergs, and falling snow. Studies show that the more immersive the experience, the more the pain relief. The VR experience seems to actually reduce pain-related brain activity and works as well as an opioid drug. Even better, the brain doesn't get habituated to VR and need larger doses of it.

If you can fly using your VR headset, be on the beach in Thailand, or even go to the moon, you're taken away from your discomfort during that time. Similarly, if your health simply prevents you from ever visiting someplace anywhere in the world, you can still go there in virtual reality for practically no cost. Personally, it is risky for me to go to countries with a high prevalence of tuberculosis (exposure is contraindicated by my life-saving medication) or where even stronger, healthier people get sick from the water. But with VR, I'll be able to tour the world.

VR can be helpful for understanding any medical condition you might have. You could put on a VR headset and "see" a medication traveling through your system or even have the immersive experience of being the medication. You can use VR to see how a medical procedure works. You could see what the camera sees in laparoscopic surgery or a colonoscopy, for instance. VR can be used to gamify learning. While the video game called Re-Mission has been on the market for years now (it helps young people with cancer understand their disease and what they need to do to stay in remission), imagine turning that learning game into a VR game. The game as a VR game could immerse the player inside the body to attack cancer using weapons like chemotherapy. Studies show it works to reinforce positive attitudes and leads to better treatment adherence.

Second Life has been used to help people lose weight and keep it off. If you're overweight, you can create an avatar—a virtual representation of yourself—within the Second Life VR environment. The avatar looks like you and moves like you if you were to lose weight. Within the VR setting, you believe it is you, but an improved version. Subconsciously, you strive toward becoming your avatar. All sorts of positive things start to happen in your brain. It's a massive motivator, and you don't even realize it.

People on the autism spectrum have been able to improve their communication and social interaction abilities by practicing them on Second Life. The patient and a therapist both create avatars and enter into a carefully controlled VR environment to work on specific issue or scenarios. The therapeutic potential for the world of autism and psychiatry in general is very large.

We're starting to see virtual reality in neurosurgery planning. The Oculus Rift can be integrated with a surgical navigation device such as the Surgical Navigation Advanced Platform (SNAP), which provides advanced imaging capabilities, including multiple points of view. Even more amazing is SuRgical Planner (SRP), an imaging platform that uses CT and MRI scans to construct dynamic, interactive models of the brain. The surgeon can plan and rehearse a surgery and see in advance what will happen. These platforms will improve the precision of surgery and give better outcomes. They'll decrease surgical time because the physician has already done the surgery through VR before the patient was ever touched.

Chapter Four

THE POWER OF THE CROWD

THE POWER OF THE CROWD—CROWDSOURCING—MEANS that you can access the massive, ubiquitous connectivity we now have to get help from very large numbers of people. Crowdsourcing means that because most people, especially in the US, have access to mobile devices and the Internet, we can aggregate the opinions, thoughts, ideas, and contributions of many, many people.

Because everyone has access to the same platform, the power of the crowd can be put to work. You can have 10,000 people doing five minutes of work each instead of five people working for seven years on the same project. The power of the crowd means we can survey or get help immediately at a mass scale for free or for a fraction of the normal cost.

In healthcare, crowdsourcing is a very powerful tool. For example, crowdsourcing allows for peer-to-peer networks for people with a specific disease. They can talk to each other and share information and ideas on an unprecedented scale. Another way of using the power of the crowd is by opening the source code of a platform so that any app developer has access to build on it. Apple's iWatch platform is open to third-party apps, for example. What that means for the consumer is that we will have many, many more apps to choose from and use, as millions of app developers will create apps for the iWatch so that Apple doesn't need to hire app developers. The iWatch platform gets populated with apps for free!

An amazing example of how the Power of the Crowd can work for patient empowerment is a company called Crohnology, founded by Sean Ahrens. Sean himself has Crohn's disease and started Crohnology as an online support community, a sort of Facebook for patients who suffer from inflammatory bowel disease. He wanted to build a patient-centered information-sharing network.

It has now morphed into a patient-driven research platform because the people who participated shared information from their personal experience and came up with insights not found in the medical literature—information that most GI doctors had no idea existed. The patients on Crohnology combined their experience, reported symptoms, rated treatments and efficacy, shared how they were feeling, and discussed what made their condition better or worse. Through this open sharing by crowdsourcing of individual patient data, it was discovered that the majority of patients with IBD cannot tolerate beer. That sort of insight is what we've come to expect from crowdsourcing, and in this case, it will help a lot of people avoid beer and unknowingly making their symptoms worse.

I met Sean for the first time in 2010 at a Quantified Self meetup. Since then, we've spoken together a few times on stage. Chronology is his first vertical; he hopes to use the same platform for some other chronic diseases.

Smart Patients takes Crohnology to another level. Cofounded by Dr. Roni Zeiger (formally of Google Health) and Dr. Gilles Frydman, Smart Patients is an online community for motivated patients and their loved ones, regardless of their condition. The community lets you learn about your condition at your own level, follow scientific developments, find clinical trials, share your questions and concerns with other members, and use what you learn in the context of your own life. When patients have the right tools, they become experts in their condition, and that knowledge improves the care they receive.

Crowd sourcing, or what I like to call CrowdThink, works in patient-centered networks because patients have a strong vested interest in talking to each other, and most importantly, getting better. Your doctor might be a specialist in your disease, but how often do you get to talk to him or her? Maybe 15 minutes every few months, or maybe even less! The other people who are experts in your disease? The patients. Not only are you going to learn about what works for them, like avoiding certain foods, you'll hear from them about what therapies might be in the pipeline, what drug combinations seem to working, where the latest research stands, and more. Sometimes the patients can be a better resource than physicians.

You do have to be cautious, however. There's no filter on these patient communities, so you have to do it yourself. Within the crowd, there will be individuals who make claims that may not be true. Someone might say that coconut oil really helped her. Maybe it did, but until a lot of other people in the crowd say the same thing, that person is an outlier and the claim needs to be treated very carefully. But when 98 out of 100 patients say it helped them too, then maybe it's time to give it a try.

The power of the crowd, especially in peer-to-peer social networking sites, provide a wonderful support system for people with chronic diseases. That's hard to find. It's very lonely being a patient. Your friends and family may be very supportive, but they don't have the personal experience of having your disease the way your social network peers do. After a while, you don't want to keep talking about how you feel with the people close to you. It starts to feel like whining; you start to feel like all you do is complain. But in your peer-to-peer group, you have a network of caring individuals who really understand and are willing to listen because they know you'll do the same for them. When you're having a difficult day or you're in the hospital or you're trying a new medication, they're there for you. They actually get it, with emotional support and also useful information about that new medication based on their knowledge and experience. And when you're in remission, or your new meds are working, they'll celebrate with you. They'll also share the information about your experience with the new drugs in the hope of helping others. It's a massive win-win.

Peer-to-peer platforms and crowdsourcing can also help build momentum to support research. Many people with rare or orphan diseases have started crowdsourced fundraising to pay for research. They've started platforms to bring patients, families, doctors, and researchers together to share insight and support. With rare diseases, insight emerges when you aggregate patient information that has never been brought together before. Treatments can be compared, and research can be shared. Patients and their families finally have someone to talk to.

I love hanging out with the high-energy CEO Jamie Heywood. He founded a couple of companies, including PatientsLikeMe, after his brother was diagnosed with a terminal illness (ALS). The amazing, and controversial, thing about PatientsLikeMe is that patients get to control and have access to their healthcare information and compare it to other patients just like them. AstraZeneca is now working with the data from Patients-LikeMe to help improve patient medication outcomes in multiple therapeutic areas.

MEDICAL DETECTIVES

We're now seeing some really interesting developments in peer-to-peer networks for diagnostics. Crowd sourcing lets people be medical detectives. CrowdMed, for instance, brings together a community of medical solvers to diagnose difficult or very complex cases. A rare medical condition could take months or even years to diagnose. Mine isn't even rare, and it took 13 years to diagnose correctly! But when you bring together thousands of experts to look at the case, an accurate diagnosis could finally be made.

Jared Hyman, the founder of CrowdMed, had the idea to harness the wisdom of the crowd to solve difficult medical cases online, many times less expensively and more quickly than the traditional medical system. On CrowdMed, patients with an unsolved medical mystery post their cases by answering a comprehensive set of medical questions, uploading relevant diagnostic and imaging test results, and selecting a subscription package. Then CrowdMed's community of over 15,000 registered medical detectives may select it and collaborate on solving it using chat, discussion, and point allocation features, while a patented prediction market technology accurately identifies the most probable diagnoses and solutions.

Using this innovative crowdsourcing methodology, CrowdMed has resolved well over 1000 real-world medical cases, with a success rate of over 60 percent, for patients who on average had been sick for seven years, seen eight doctors, and incurred over $70,000 in medical expenses before submitting their case. The average case resolution time is just two to three months, and the cost is under $500 per case. CrowdMed's amazing results have been submitted to major peer-reviewed medical journals for publication and presented at prestigious healthcare events including the PMWC, Health Innovation Summit, and TEDMED.

The beauty of CrowdMed is that it brings together the perspectives of healthcare professionals from a wide range of fields. They share their experience and cross-pollinate their ideas. When their collective wisdom is brought to bear on a medical mystery, it gets solved.

The power in numbers means the power to change things.

When people with a disease combine into a crowd, they have the power to share information, influence research funding, change laws, raise awareness, feel in charge instead of powerless, and improve their lives. It's transformative.

INDEPENDENCE THROUGH THE CROWD

Have you heard of the Spoon Theory? It's an intangible unit of measurement for patients with a debilitating disease that represents how much energy they have in a given day. Imagine that every morning you have only 20 spoons of energy. They represent all the energy you have that day to do everything you need to do—from taking a shower, to loading the dishwasher, to getting to work or school, or taking care of the kids. Some days, just taking a shower can use up five of those spoons. That leaves you with just 15 spoons, and a trip to the grocery store could take 12 of them! That leaves you with just three to do everything else you need to do that day.

The power of the crowd is changing that. Just a few years ago, only wealthy people could afford to hire all the help it takes to manage your life as a patient on your own. Now, crowd technologies have democratized having a chauffeur, having a personal shopper, having a grocery delivery service, and having, well, almost anything.

If you have access to the Internet, you've definitely heard of Uber. It's a platform that instantly connects riders to drivers. For someone like me, it's a game changer. There are times when I'm just not well enough to drive, or I need to take pain medication, but I still want to live my life and not be stuck at home. That I couldn't just hop into my car and drive somewhere whenever I wanted to used to isolate me. Public transportation isn't always an option for many reasons, including limited access and not exposing myself to a large number of people if I've just gotten my immunosuppressant medication. Sometimes I'm not feeling well and just don't want to deal with the influx of sounds, sights, noise, and people on a bus while trying to deal with pain and other symptoms. When I can't drive myself, I'm sharply reminded of how people with serious health issues are incredibly alone and often suffer from a lack of independence. Being reliant on other people for your needs can be very difficult for your emotional state.

Now, however, on days when I'm not well enough to drive, I'm no longer isolated. Where I live in Silicon Valley, Uber is easily accessible, reliable, fast, and affordable. In fact, in my area, Uber costs less than a taxi! Uber uses the power of the crowd for its drivers. You have to rate the service every single time you use it, so that crowdsourced feedback gives Uber a huge amount of data about their drivers and their riders. Uber has made headlines around the world for its innovative service. It's not without concerns about driver safety and other issues, but the Uber concept is extremely efficient and provides a badly needed customer service in many places. It has really helped open up my life and take a huge load of stress off me. I no longer constantly worry about transportation.

The Uber concept is spreading to provide just about every other service. Companies like Instacart use the power of the crowd to provide groceries delivered to your door within a couple of hours for about $4 per delivery—a cost that's accessible to almost everybody. I use it once a week, and I never worry about carrying heavy groceries, taking the time to drive to the store and shop, or coming home from a hospitalization, work, or trip to an apartment empty of food. Like Uber, Instacart uses the power of the crowd for their workforce. If you have a chronic health problem, this sort of service is hugely helpful because going to supermarket can be very exhausting. If you're immunocompromised, it's risky to expose yourself to any public space like a supermarket or drugstore. You can't be or shouldn't be in these places. With services like Instacart, you don't have to be.

Leah Busque has made a major impact by founding a company called TaskRabbit, which uses the power of the crowd as the workforce. You can post any kind of job on TaskRabbit and quickly find someone to do it for a reasonable price. I use TaskRabbit when I need help with things like getting rid of a ton of moving boxes, picking up dry cleaning or stuff at the drugstore, or getting a video transcribed. Again, for people whose health makes managing a trip to the drugstore a problem, these companies restore a lot independence. No longer do you have to wait until a friend or family member can fit your errand in. You can be proactive and get it done yourself, inexpensively, and almost immediately.

Mike Chen and Aaron Kemmer recently started a company called Magic that will do pretty much anything you want for you! All your communications are by text message. You could text them and say you need a dozen roses delivered to your mother by 3 p.m. today or you need to pick up a rental car in Miami at 5 p.m. They take it from there. Companies like these have democratized the service economy by making services much less expensive and much more accessible.

Suddenly, patients have an affordable, accessible world full of independence.

Chapter Five

ARTIFICIAL INTELLIGENCE

ARTIFICIAL INTELLIGENCE DESCRIBES COMPUTERS THAT can perceive, imagine, and reason the same way humans can. AI is sometimes called machine intelligence, machine learning, or deep learning. They all describe a really smart computer that can think for itself. An important component of AI is natural language processing, which allows the computer to understand normal spoken language, even when it's ungrammatical or colloquial. Siri, the virtual personal assistant who comes with your iPhone, is AI in your pocket. Siri uses a natural language user interface to communicate with you. The AI really kicks in when Siri learns from you, getting better at understanding your language usage and preferences every time you talk to her.

A central concept in AI is the Turing test, which evaluates a computer's ability to behave in ways that are indistinguishable from a human. A computer that can think and reason like a human will pass the Turing test. Right now, a lot of companies are putting major money into making AI that passes the Turing test into a reality...and they're getting very close. Once it passes the Turing test, AI is completely indistinguishable from a human. It will be able to hold a conversation with you. The only way you would be able to tell you're not talking to a human would be if you were sitting in front of it and could see it wasn't. And with soft robotics on the way, even that way of distinguishing might go away. In fact, there's a higher-level test called the Total Turing Test, where the AI has perceptual abilities (computer vision) and can manipulate objects through a robot. AI that can pass the harder test is on the way.

WORKING WITH WATSON

A leader in this area is IBM Watson, an artificially intelligent computer system created by IBM that can answer questions posed in normal human language. This is the computer that beat the Jeopardy champion in 2011. That was an early version; since then, Watson has become a lot smarter. IBM showed that Watson goes far beyond simply searching a database. It can also reason and learn from vast amounts of data.

The next step was to send Watson to medical school. Watson has read millions of pages of medical textbooks, as well as all the millions of medical journal articles stored on the PubMed and Medline databases. Its looked at case studies and patient histories and much more, all to tune up its ability to make fast, accurate diagnoses and decisions. Some 170,000 clinical trials

"You can't list your iPhone as your primary-care physician."

Kaamran Hafeez / New Yorker; © Condé Nast

for cancer go on in the world every year. That's a huge number! How in the world, as a physician, can you keep up with that? You can't, but Watson can. That's where AI can give a physician a huge boost. AI not only keeps up with all the studies, sorts through the noise, and analyzes them but also learns from them and can answer questions using the information in them.

The reason physicians like Watson is because it can help them get to the right diagnosis quickly. The doctor provides all the case information; Watson mines the patient's history for more data. Then it not only provides a list of the top three most likely diagnoses for the patient but it also provides the reasoning and the evidence trail behind it. It tells the doctor why it thinks a particular diagnosis is correct. For doctors, the beauty of Watson is that the diagnosis is evidence-based and includes the most current information. It's free of cognitive bias and overconfidence, both problems that can unconsciously affect a physician.

At some point, Watson may well replace a lot of what healthcare providers do. The famous VC and cofounder of Sun Microsystems Vinod Khosla talks about AI completely replacing the need for the physician. It might seem strange to let a computer diagnose your medical problem, but most people have already encountered AI in other situations, such as the customer service department when you want to talk to someone about a problem with an order. That usually works out pretty well, so it's not that much of a leap to talk to AI when you call the pharmacy with a question about a drug's side effects. And because AI never gets tired, never gets grumpy, and never sleeps, it will always answer the phone.

Personally, I think that we will always need doctors for the emotional side of medicine to provide support and insight and answer the sort of questions only a human can. But when you're talking about the human body, if you look at it just like a car or an airplane, it's an operating system and hardware, which means you're just data. We haven't completely mapped out and decoded everything yet, but we will. Everything about you can be seen just as terabytes of data points.

The average person will generate one million gigabytes of health-related data across a lifetime. That's as much data as 300 million books. No human being can fully understand the human body, but AI can analyze you in seconds. AI can take the massive data set that's you and discover things about you that you could never know otherwise. AI can assess you and diagnose you and give you a better outcome. And the amazing thing is that you can access all that computing power through your smart phone. We hold these massive computers in our hands 24/7 now. Our phones are the first thing we look at in the morning and the last thing we look at when we go to bed. We've got AI right here. It's accessible to us. It doesn't need some kind of distribution channel because it's already distributed. The technology in our hand is catching up to AI.

Companies like Google are putting massive amounts of capital into AI. Ray Kurzweil, the amazing inventor, computer scientist, and futurist, is now a Director of Engineering at Google with a huge budget for computer processing. Google is hiring the best AI people out there, as well as acquiring companies like Deep Mind, who are making some major breakthroughs in AI. Watch Google in this space to follow some of the fast moving innovation in AI.

AI is accelerating even more, as IBM has opened the application programming interface (API) for Watson for anyone to access. It doesn't matter who you are—a Fortune 500 company, a two-person start-up, a hospital system—you now have access to Watson. Why would IBM do that? Because AI improves the more it's used. If IBM can get millions of people using Watson and experimenting and researching with it, IBM will get that much more successful that much more quickly. They will beat out all the rest of the competition. And AI that passes the Turing test will become a reality.

IBM Watson is working on a lot of interesting health applications, so many that I can only highlight a few here. One is a joint venture with Talkspace, an app that connects you to a skilled mental health professional for online therapy via text. Some 45 million people a year are diagnosed with mental health issues each year, so there's a great need for an inexpensive, accessible way to get trained help. The hard part is matching you up with the right therapist, someone who has experience with your particular needs and will be compatible with you. That's where AI comes in. Talkspace has partnered with IBM to use Watson's Personality Insights API has a way to understand the mental state of Talkspace users. The API breaks down and analyzes the verbal cues in your texts to pair you up with a therapist who will be a good match for you.

IBM is also partnering with massive healthcare companies like Johnson & Johnson and Medtronic to find ways to combine AI with medical technology. With Medtronic, they're taking advantage of the Internet of Things around medical devices. They can collect and anonymize data about their heart pacemakers or insulin pumps and get a better understanding of how well they're working without compromising patient privacy.

IBM has struck a deal with Apple to collect information from wearable devices linked to Apple HealthKit. The platform will aggregate all of a patient's health data, whether it's from wearable tech devices, EMRs, or even eventually from the environment.

Medtronic, a maker of implantable heart devices and diabetes products, will use Watson to create an Internet of Things around its medical devices. They'll collect data to give to individual patients, then anonymize the data for Watson to analyze. The idea is to understand how well the implants are working over time. Similarly, Watson Health Cloud uses anonymized data from healthcare companies likes Phytel and Explorys (both of which IBM acquired) to identify trends and answer natural language healthcare queries. And IBM's latest partnership with CVS Health seeks to identify ill individuals who would benefit from proactive healthcare. Watson would suggest coordinated cost-effective primary and outpatient care options based on a more holistic view of each patient.

Researchers are now experimenting with replicating brain structures outside the human body. We're looking at electronic cells that can store information but also help create AI. Once we're able to store and remember and recall past events using a cell that we've grown, we can start developing it as a storage component for an AI network. We'll essentially be able to replicate the human brain and put it inside a robot.

Other AI applications include things like augmented reality glasses to help visually handicapped people. For people who still have some sight, as most visually handicapped people do, AI can improve their independence or even allow them to be independent for the first time. The glasses would be small, comfortable, and unobtrusive. They could restore depth perception and color vision. Service dogs and service ponies could become a thing of the past!

Google Glass is probably the best-known version of augmented reality glasses. Google Glass was introduced in 2012 with a lot of fanfare and publicity. As a consumer product, it never really caught on. Google withdrew it from the consumer market in 2015. In the healthcare market, however, Google Glass is alive and doing very well.

Surgeons are using Google Glass in a variety of innovative ways, such as monitoring a patient's vital signs without having to look up at a monitor. It's also being used in medical training to stream the surgeon's viewpoint during an operation. The video of the operation can become part of the patient's medical record. Glass allows emergency responders to stream images of a patient from the field and during transport; the emergency room team is then better prepared to act quickly when the patient arrives.

Google has released the code for Glass, so now there are a lot of people working on applications. One is Augmedix, an app that helps doctors do their charting more efficiently. The doctor wears Google Glass during the office visit and has a normal conversation with the patient—without a computer between them. Augmedix sends the audio-video feed from Google Glass to a remote, HIPAA secure location, where a trained scribe uses it to enter patient notes in the patient's electronic health record. That frees up time for the doctor to concentrate fully on the patient during a visit and cuts down on time spent entering information into the patient's record. The doctor later reviews the notes and signs off on the chart. Doctors today spend about a third of their time in front of a computer just entering data. Augmedix can save them hours of routine charting work each week and give them more time to spend with patients.

Arthur C. Clark, author of many classic science fiction books, once said, "Any sufficiently advanced technology is indistinguishable from magic." Researchers are working on AI algorithms with the capacity not only for logic and natural conversation but even flirtation. That's pretty mind-blowing because it means AI can understand manipulation. By encoding thought vectors, AI systems could eventually have human-like common sense. That's both hugely encouraging and very frightening. It raises a lot of difficult questions. At what point should you be made aware that you're talking to a computer? How can we put safeguards in place to make sure AI stays within our control?

IBM Watson is now doing cognitive cooking, inventing unique recipes that combine ingredients and flavors in new ways. Watson takes its existing knowledge of what ingredients and flavors combine well, and then extrapolates and looks for new patterns. Of course, Watson can't taste anything, so humans (or robots!) have to get involved not just to prepare the dishes but taste them. Exactly the same approach of extrapolating and looking for patterns can be applied to the human genome in the search for disease and treatments. Now that genome sequencing is so inexpensive, a huge new area for AI analysis has opened up. Right now, analyzing the sequence is done just with brute force computing power, which takes a long time. When you analyze with AI instead, the picture changes.

Another valuable use for AI is in pharmacology. Thousands of drugs are already on the market, and new ones become available all the time. It's very hard for a doctor to keep up with latest drugs even in a subspecialty. With AI, prescribing the right drug for an individual patient, in the right dose and without interactions with other drugs, becomes much easier and safer. AI will prevent a lot of inadvertent drug overdoses and bad reactions. AI will also find new uses for old drugs. Remicade (infliximab), a drug I take for Crohn's disease, was originally approved to treat rheumatoid arthritis. It's now used for other autoimmune conditions that seem to have a similar underlying genetic cause and involve tumor necrosis factor (TNF), including psoriatic arthritis and ankylosing spondylitis. Doctors have always used drugs off-label with mixed success. With AI, drugs created to treat one condition can be studied to see if they'd be good candidates to treat additional conditions. The process would move far faster than the trial-and-error approach that's taken now. Expect big surprises.

AUTONOMOUS AUTOS

Autonomous cars are another awesome area for AI. These electric-powered vehicles will drive themselves. The advantages are huge in terms of carbon footprint, improved traffic flow, considerably fewer accidents, and personal convenience. Instead of watching the road on your daily commute, you could read or sleep. Road safety will improve. You can't fall asleep at the wheel or be distracted or drive drunk in an autonomous car. For people who can't drive now for whatever reason, autonomous cars would give them the same personal mobility every one else has. There are still some barriers to entry, such as human interaction, but these barriers will fall, probably soon, because testing and improving are moving ahead very quickly.

The University of Michigan has built MCity, a 32-acre environment for testing self-driving cars, on its campus in Ann Arbor. By 2020, estimates are that about $20 billion will be spent on adding sensors for self-driving cars to smart cities. The U.S. Department of Transportation has issued a $42 million grant to have cities make their vehicles "smart" and able to talk to one another. New York, Tampa, and Wyoming are all participating. New York is installing the car-to-car technology in 10,000 cars, buses, and other passenger vehicles that are run by the city. These cities will also be adding the tech to parts of the infrastructure, like traffic lights. New York is hoping to cut unimpaired traffic accidents by 80 percent.

Of course, much safety testing remains to be done, but I believe that by 2030, self-driving cars will be widely adopted across the country. Within a few years of that happening, we'll start to see laws around even being allowed to drive manually, or needing a special license to drive. The cost to add the autonomous feature won't be significantly higher than for a regular car. Cars can then have the opportunity for a whole new redesign, as the driver's seat won't be needed. Maybe they will be redesigned with four seats facing each other, more like a living room or meeting space.

The implications of greater road safety are interesting. Safer is always better, but the Office of Science and Technology Policy at the White House points out an upcoming problem we will be facing: greater road safety means fewer organs available for transplant because we get a lot of donated organs from people who die in car crashes. Fortunately, as I'll explain in chapter 7, in the future we'll be able to make organs instead of relying on donors. I'm interacting with the OSTP with two companies, i4j (Innovation for Jobs), and the Organ Preservation Alliance. I find the OSTP staff are impressive with their forward thinking. Since they exist to inform and educate the Executive Office of the US (including the President), I am extremely grateful I get to work with them to help improve people's lives.

OTHER AI APPLICATIONS

We're also starting to see other applications of AI in our day-to-day lives. The Isabel Symptom Checker uses AI to diagnose your health problem. It uses natural language to analyze the pattern of the symptoms you enter and decide on the most likely diagnosis based on a huge medical database. Isabel is used by a lot of organizations such as hospitals and health insurers. You can also download it as a free app.

Amazon Echo is a voice command device that's essentially a small computer. It looks like a tiny upright cylinder fan. Echo analyzes everything around you. It's hands-free, always on, has seven microphones, and uses beam-forming wireless technology. It can hear you from across the room even while music is playing. Echo connects to Alexa, a voice-activated service in the cloud that instantly provides information, answers questions, plays music, reads the news, checks sports, weather, and so on. You can be across the room and say, "Echo, what's the weather going to be tomorrow?" Your voice is picked up and sent to the AI in the cloud where it is analyzed using natural language processing, and then Alexa tells you the answer. Alexa doesn't just assemble information; it analyzes it and draws conclusions.

For example, while staying with Taylor Milsal, she asked her Echo "Who is Robin Farmanfarmaian?" Alexa immediately came back with "Robin Starbuck Soloway Farmanfarmaian is an entrepreneur, technologist, patient advocate, medical futurist, and socialite." Most of that information is easy to find from my LinkedIn profile and all the many online "abouts" that include my name. But the "socialite" description is different. I'm not described anywhere on the Internet as a "socialite." However, I was a president and gala chair at the San Francisco Ballet for a number of years, and I've done a lot of charity work. Photos of me attending various society events have been published many times in newspapers and magazines. The implication? Alexa is analyzing my Internet appearances. It's looking at the types of coverage I've received in magazines and newspapers and not just reciting information from top Google search results.

Echo does raise some disturbing privacy questions because it can be seen essentially as a spy recording everything you do! But it's also very cool and costs under $200. Imagine how it could help someone with Alzheimer's disease, for instance. Echo could answer questions like "What's my name?" over and over again without ever getting irritated or tired, as a human caregiver might.

In fact, AI is great for patient companionship. The Pepper humanoid robot from SoftBank Mobile went on sale in Japan in June 2015 for $1,600 and sold out in one minute. Pepper can analyze emotions using voice recognition and a bank of cameras and sensors. It makes gestures and can even dance, but Pepper isn't meant to perform any tasks. Instead, it uses AI to be your emotional companion. It literally gets to know you, responds to your emotions, and makes you feel happy. Pepper wants to be your friend, and he is adorable. OK, maybe a little creepy, but an emotional robot could be very helpful for people isolated by disabilities or dementia. Or for those of us who talk to ourselves…. And perhaps in the near future Pepper will notice if you've stopped responding or are responding strangely and call for help. That technology application is already being successfully developed and just needs to converge with Pepper.

If you find Pepper a little disturbing, you may find the idea of a robotic sex companion even weirder. Do you want to have such an emotional connection with a robot? Why not? My friend Dan Barry, M.D., Ph.D., is a former astronaut and a world-leading robotics inventor. His company, Denbar Robotics, specializes in robotic systems for healthcare settings and in assistive devices for people with disabilities. He points out that millions of people can't have sex with another human for various reasons. They're cut off from that aspect of life, but emotional robots could help, as could eventually robots for sex. Disclaimer: Dan's company is *not* working on sex robots. He's a robotics expert who can easily imagine a world where this happens.

There are different ways to look at that, but good or bad, it doesn't matter. It's happening. There's no way to stop the future of AI, including stuff we haven't even thought of yet. AI will only get faster and better to the point where it's going to greatly surpass human intelligence. What amazing insights will that give us!

Chapter Six

ROBOTICS

WE CAN LOOK AT THE AUTOMATED MACHINES WE CALL robots in a variety of ways. Some robots are in, around, or on your body. They're interacting with you personally and physically by touching you in some way. Some robots are automated or are controlled by people, such as assembly line robots and surgical robots. And some robots use AI to take the place of a human in some respect. All these categories will have a massive impact on our lives.

EXOSKELETON ROBOTS: AUGMENTED MOBILITY

A great example of a wearable robot is the Ekso Bionics exoskeleton wearable bionic suit, designed to help people with paralyzed legs learn to stand and walk again. The Ekso is an exoskeleton—an external framework fitted with sensors and battery-powered motors that drive the legs. The sensors detect a shift in the patient's weight and kick in to activate the motors, which help lift them up out of seat or wheelchair. Once the patient is standing, the sensors and motors shift to taking steps. Most patients learn to stand up and start walking with a walker or crutches in the first session. Most of today's suits weigh between 22 and 50 pounds and cost more than $70,000, but I expect the price to fall significantly over the next decade in the same way the price falls for other tech advancements.

The Ekso Bionics device is designed for rehab work with stroke and paralysis patients. The Department of Defense is very interested in exoskeletons, not just because they help injured soldiers walk again. It's because exoskeletons essentially let you put yourself inside a robot. The ability of an exoskeleton to augment human strength, endurance, and mobility has very obvious military uses.

DARPA, which stands for Defense Advanced Research Projects Agency, has put a huge amount of money into exoskeleton research. DARPA is responsible for developing cutting edge military technology. Looking at what DARPA is working on is a good way to see what disruptive and game changing technology is in the pipeline. DARPA famously invented the Internet and is behind almost every platform we use now. 80 percent of the components of your smart phone were developed by DARPA.

A lot of advanced military technology eventually finds its way to civilian use. The exoskeletons being developed today may someday be routinely worn by people who are paralyzed or have degenerative neuromuscular diseases. In fact, anyone who has trouble moving around for any reason could someday use an exoskeleton to regain independence. Being stuck in a wheelchair because you're paralyzed from the waist down could become a thing of the past. No longer will you need to rely on other people for basic needs. No longer will you have to look up at everyone and ask for help to get up out of your wheelchair. Exoskeletons will restore your dignity and independence.

We're already seeing augmented mobility exoskeleton suits being used by some workers who have to do a lot of lifting. In Japan, some workers, such as airline luggage handlers and bank currency movers, are wearing robotic exoskeletons called HAL (hybrid assistive limb) that pick up bioelectric signals from the person's muscles and send them to the exoskeleton to augment their strength and movement. Given that 25 percent of Japan's population is over age 65, this is a great way to help older workers stay on the job. This same technology can be used to help older patients with mobility problems or augment healthcare workers so they can more easily lift patients. The Japanese exoskeleton suits lease for about $2,000 a year. They weigh about 50 pounds, and the battery lasts a little over two and a half hours. When used to lift things, the suit reduces the perceived weight of an object by 40 percent so that a 20 pound weight feels as if it weighs only 12 pounds.

ATTACHABLE ROBOTS

Exoskeletons fit around you on the outside. What about robots that attach to you? In 2014, Dean Kamen, inventor of the Segway, received FDA approval for his robotic arm. Designed for amputees, the DEKA Arm System is incredibly advanced.

Wearers can do things that are impossible with more traditional prostheses.

Here's how it works: The DEKA arm contains a complex set of switches, movement sensors, and force sensors that enable it to move in many different directions with varying levels of force. The DEKA arm can perform complex and delicate tasks, like picking up a grape, but it is also strong enough to hold a power tool. The arm is set into motion by electrodes that are attached to the wearer's muscles in the remaining part of the arm. The electrodes detect the electrical activity produced when those muscles contract. They transmit the signal to a computer processor in the DEKA arm; the computer then translates those signals into ten different types of movements.

A few years ago, I got to visit Dean Kamen's eclectic, hexagonal mansion in Bedford, New Hampshire. It was amazing! He has a helicopter pad and inventions all over the house. Dean has more than 150 patents for innovative products and technologies, including the Slingshot—a water purification system the size of a small fridge that only costs pennies per gallon to purify. He partnered with Coca-Cola to distribute them all over the developing world. In 1989, Dean started FIRST, an acronym for Inspiration and Recognition of Science and Technology. The FIRST Robotics competition has spurred on innovation in robotics, encouraging and rewarding thousands of teams made up of high school students.

At some point in the future, the interface between the processor in the prosthetic limb and the body won't be electrodes on the skin; it will be a direct brain-computer interface (BCI). You'll just think about moving the limb, and it will do exactly what you want. You'll be able to feel the response because artificial skin will cover the robotic arm and transmit all the sensory information skin does, like temperature and texture. Because prosthetics today are essentially numb, their usefulness is limited, as is the willingness of users to wear them.

Robotic limbs could move beyond enabling amputees to super-enabling them. The prosthetic limb could be stronger than a human limb. It could have a range of motion far beyond what a human has. I can't rotate my wrist through 360 degrees, but a robot arm could do that. Wouldn't it be great if you could design an arm from scratch? What modifications would you make? An additional thumb? An elbow that rotates 360 degrees? Two hands on one arm? With 3-D printing converging with robotics and BCIs, we will only be limited by our imagination.

SURGICAL ROBOTS

Robotic surgery uses tiny instruments attached to a robotic arm to perform minimally invasive surgery (surgery without large incisions) on patients. These instruments are controlled directly by a surgeon through a computer interface that translates every motion the surgeon makes with their hands outside the body into an identical motion of the very small surgical instrument inside the body.

Minimally invasive surgery has been around since the 1990s primarily as laparoscopic surgery, in which the surgeon uses endoscopes (tiny cameras) and long, small-diameter instruments inserted through small incisions in the body. It is widely used for fairly simple operations. The small incisions, often less than half an inch in size, mean less pain, smaller scars, faster healing, and a shorter hospital stay for the patient than an open surgical incision, which is big enough to get the surgeon's hands inside the patient.

Robotic surgery builds on the basics of laparoscopic surgery, but it is even more precise because the instruments are controlled by the computer, not the surgeon's hand. Probably the best-known surgical robot is the da Vinci Surgical System, made by Intuitive Surgical. Strictly speaking, the da Vinci isn't a surgical robot because it's not programmed and can't make any decisions on its own. It just mimics the surgeon's hand motions. It's more like a computer interface between the surgeon and the micro instruments that do the actual work within the patient's body. It's pretty close to a robot, however, and the term robotic surgery is now widely used.

My friend Dr. Catherine Mohr is head of research at Intuitive Surgical. She's a great person to follow online and at conferences for updates on surgical robotics. She says it is like the da Vinci gives the surgeon "superpowers," such as better eyesight and dexterity. The da Vinci Surgical System's monitors gives a magnified view inside the patient's body in 3-D and HD. The instruments the surgeon manipulates can bend and rotate far more than the human wrist can, and the movements can be scaled. That means the surgeon's hand movements get translated into smaller, more precise movements of the tiny instruments inside the patient. Unlike laparoscopic surgery, the surgeon isn't in the operating room standing beside the patient. Instead, he or she sits at a console in a room nearby, using the console monitors to see what's going on and using the controllers to move the instruments. A special cart with three or four robotic arms is positioned next to the patient. The arms can only move if the surgeon tells them to with the controller. The instruments at the end of the arms each have a very specific function, such as clamping or suturing.

The da Vinci is game changing technology. It's been used successfully worldwide in more than 2.5 million surgical procedures. The da Vinci is changing the way we do surgery, but like an MRI or other large pieces of equipment, it is very expensive initially for the hospital to buy one, and it requires extra training for the surgical team. The da Vinci is also a bit controversial because the equipment costs are more than the equivalent surgery using more conventional (open or laparoscopic) methods. The costs are offset in many cases at least in part by shorter hospital stays and lower rates of readmission to the hospital. Some studies suggest that the outcomes from robotic surgery aren't any better or different than the outcomes from standard laparoscopic surgery, but in most cases, it replaces open surgery for surgeries that are too complicated to do laparoscopically. It extends the number of surgeries that can be done with minimal invasiveness to include things like my bowel surgery. We'll probably get resolution on that issue as the da Vinci system keeps improving and becomes even more accurate and capable of even more precise operations.

In a recent demonstration of the da Vinci's capabilities, it was used to reattach a piece of the skin of a grape using sutures—through the narrow neck of a glass bottle. That's a pretty delicate operation that no human could do through such a small incision because no human can get a hand inside the bottle. The da Vinci can perform miniaturized tasks, such as tying the incredibly tiny knots used to reattach the grape skin, inside a confined space.

I've actually been allowed to play with a da Vinci at Intuitive Surgical's headquarters. Not on anything alive, of course, but I was able to get a good sense of how intuitive it is to use. It takes away any tremor in the surgeon's hand, and the robotic arms don't get tired.

I wish the da Vinci was available to me years ago. I've had surgery where they've opened me up 12 inches, and I've had surgery that was laparoscopic with and opening only half an inch. When I had major surgery to remove organs, I ended up in the hospital for eight days and spent weeks recovering. When I had my gallbladder removed laparoscopically later on, it was outpatient surgery that only took about two hours. A surgical robot is making it possible to make all surgeries like my gallbladder operation.

While da Vinci still needs a human operator to perform the surgery, think what will happen when an AI component is added. Within the next 20 to 25 years, an AI doctor could diagnose your medical problem, and an AI robotic surgeon could do the operation. The da Vinci system is already being used for telemedicine, but only for teaching purposes. Surgeons who want to learn how to use da Vinci can work with specialists around the world without having to leave their home hospital. But if you can operate from across the room now, in the future it might be possible for a doctor in San Francisco to operate on someone in Europe. In fact, in 2001, a team of physicians in New York made history by operating remotely on a patient in Strasbourg, France, via a surgical robot named Zeus. Their success depended on broadband transmission capability with optimized compression that limited the time delay between the doctors' commands and the robot's response.

ROBOTS IN YOUR BODY

One way to get a robot into your body is to swallow it. You can do this now with a swallowable capsule that does a virtual colonoscopy. PillCam COLON, approved by the FDA in 2014, uses a miniaturized camera contained in a disposable capsule that naturally passes through the digestive system. The capsule, about the size of a vitamin pill, contains a small robot camera that travels through your GI tract and takes pictures as it passes through your colon. The pictures are wirelessly transmitted to sensors and a recorder you wear on a belt around your waist.

On the day of the exam, you go to the doctor's office to have the sensors placed and swallow the capsule. You then take a laxative and go home or back to work. The pill works its way naturally through your system—it can take as long as ten hours—and sends a stream of images to the sensors. The pill passes out of your body the usual way. Fortunately for patients, it's disposable, so you don't have to retrieve it! The sensors and recorder get returned to the doctor, who analyzes the images.

This is great for patients because a virtual colonoscopy is much more convenient and less unpleasant than a physical colonoscopy. Instead of being anesthetized for the physical colonoscopy, you can just swallow the capsule. You still have to do some bowel prep, but the process is much easier. For people who can't have a traditional colonoscopy because of anatomic problems in the colon or some other reason (they're not recommended for people over age 80), the camera capsule is a good alternative.

Current guidelines say adults over age 50 should have a colonoscopy to screen for colorectal cancer at least once every ten years, and more often if you have risk factors. The idea is to catch colon cancer early because it's very treatable in the early stages. Does this happen? No, because having a colonoscopy is so unpleasant and time-consuming that people tend to skip it. Because the PillCam is so much easier, we can hope that in the future more people will get screened. This one technological innovation has the potential to save millions of lives.

The PillCam isn't just for colon cancer screening, however. The PillCam ESO can be used for visualizing the esophagus. Capsule endoscopy with the PillCam SB can be used to image the small intestine, which is a much more difficult procedure than a colonoscopy. As a Crohn's disease patient, I'm very interested in this. Many Crohn's disease patients have lesions in the small intestine, but these are hard to detect. The PillCam makes it much easier, which makes managing the disease better. If you know you have lesions in the small intestine, you might need to change medications, for example.

MICROROBOTICS

Remember the movie *Fantastic Voyage* from way back in 1966? In that film, a submarine with its crew was somehow shrunk to microscopic size and injected into the bloodstream of an injured scientist to repair damage to his brain. In addition to featuring Raquel Welsh in a clingy body suit, the film had great special effects showing the sub passing through various parts of the body.

Miniaturized tools that move through the bloodstream to repair damage was science fiction then, but they may be on the verge of becoming reality. Instead of a mini sub with a crew, a swarm of microrobots could be injected into the body to deliver a highly targeted dose of a drug or radioactive seeds to treat cancer, to clear a blood clot, to perform a tissue biopsy, or to build a scaffold around an area where new cells need to be grown.

A swarm of microrobots would be very, very precise when it came to things like delivering drugs directly to a tumor. That would be a huge improvement on injecting the drug into the bloodstream and damaging noncancerous cells along with the cancerous ones. Microbots could help with lots of other problems as well. Right now, drugs that treat the retina are injected into the eye, where they slowly diffuse. Only a fraction of the drug actually reaches the target. Microbots laden with drugs could potentially deliver them in a more targeted manner. That would give better results with a much smaller dose and fewer side effects and unwanted interactions.

Heart attacks and strokes happen when blood clots block arteries. If you're having a heart attack, the clot can be removed by inserting a catheter with a balloon tip into your arm or groin, snaking it up to the blockage, and inflating the balloon to open the artery. This procedure could be replaced by one that simply injects a swarm of microbots that go to the area and drill out the blockage. Right now, treating a blocked artery in the brain is difficult—catheters don't work very well and drugs to break up the clot are risky. Microbot drills could really revolutionize stroke treatment.

Microrobots could also help with disease diagnosis. At Johns Hopkins University, researchers have developed microgrippers, star-shaped devices that measure less than 500 micrometers from tip to tip—about the size of a speck of dust. The microgrippers are made of materials that respond to environmental factors, like temperature, pH, and enzymes. The temperature-sensitive microgripper arms will close when exposed to the body's heat, for example. After being put in the correct place using a standard endoscopy tool, the arms close around a tiny piece of tissue to do a miniature biopsy. Each microgripper also contains a tiny bit of magnetic material, so a magnetic tool is used to retrieve them.

Right now, biopsies are an invasive procedure using needles or scalpels to get the tissue sample for examination under the microscope. They're painful and open you up to the risk of infection, and sometime they miss the right area. A swarm of microbots with grippers could do basic biopsies by taking a lot of much smaller tissue samples. Statistically, that's more accurate than taking just a few larger samples from the suspect area. Even though a lot of microgrippers will need to be used, the experience will be much easier for the patient with very little pain and better results.

Microfish, 3-D printed microrobots that swim through the patient's bloodstream, could in the future be used to deliver drugs or remove toxins. The microfish will contain functional nanoparticles that will let them be self-propelled by chemical energy; they'll be steered from outside the body by magnet.

Microrobotics still face a lot of engineering obstacles. At the microscopic scale, every aspect of robotic operations needs to be completely rethought and redone. Power and movement become especially tricky. The human body has many levels of constraints that you wouldn't normally see when you're talking about big robotics. You need to keep track of where the robot or the swarm is inside the body. You have to make sure it's not toxic, that it won't injure tissues, and design it so that it can leave the body safely once it's done what it's supposed to do.

One of the biggest problems for miniaturizing robots is the power source. When you shrink an object down below one millimeter, how do you get a battery inside of it? Compared to the rest of miniaturized technologies, battery power is an area where we haven't been able to make much headway. It's just not moving as quickly as other technologies. A lot of money is now being attracted to batteries, however, so we'll probably see some breakthroughs there in the near future.

It's possible to send radio waves into your body to generate electricity, but it's also difficult to do. In that case, to be able to harvest the electricity the microbot needs some kind of an antenna. It can't be too small or else it won't be able to collect a meaningful amount of energy. That really limits how small the microbot can be.

One alternative to battery power for microbots is using chemical power. The ingestible sensor-enabled pill from Proteus Digital Health that I talked about back in chapter 1 gets its power from the body's own gastric acid. In the future, we could use biohybrid microrobots, where living bacteria are harnessed to do the swimming or even pursue a target based on chemical signals. Magnets can also be used to move the microbots from outside the body.

Precise placement of microrobots is another big issue. It's crucial to get them to exactly the right place, but that's difficult. Even the most sophisticated microswimmer may not be able to combat the powerful currents in the bloodstream and navigate to where it's needed. An external guidance system could be used to help the device figure out its way to its destination. One obvious way would be to incorporate a magnetic material into the microbot and use external magnets to steer it. The goal is to have the microbots roaming freely with no connection to the outside of the body, but in the interim, maybe the microbots could be tethered to catheters that are threaded into the blood vessels.

DRONES

Drones are unmanned aerial vehicles that can fly autonomously, without direct control by humans. Their potential for changing how we deliver healthcare is huge and already underway.

Automatic external defibrillators (AEDs) are already common in public places, like schools and shopping malls. An AED is used to save someone in cardiac arrest, but it only helps if it's applied very quickly. If you collapse in a place that doesn't have an AED, or if the AED doesn't get to you quickly enough, you might not survive. What if someone saw you collapse and called 911 and instead of sending an ambulance, they sent an ambulance drone? This would be a small flying machine with a built-in AED, a video screen that gives directions on how to use it, and a webcam link to an emergency room doctor. It would fly directly to your coordinates at about 60 miles an hour without having to fight traffic, and it could reach you in under a minute. Survival rates for cardiac arrest could go up dramatically. Prototypes of flying AEDs are already being tested.

Nick Soloway (aka my dad) has just started a new company that makes inexpensive defibrillators that are designed to be used by anyone, essentially, defibrillators for dummies. Current AEDs are complicated and expensive. Someone who's never seen one before wouldn't be able to use it effectively within the short time frame needed to save someone in cardiac arrest. Having a drone deliver one of these defibrillators, which anyone can use correctly right away, would be very valuable.

In Europe, nearly a million people experience cardiac arrest every year, but only 8 percent survive because of the slow response time for emergency services. That"s huge. Flying AEDs could increase the survival rate from 8 percent to 80 percent\.

There's a problem with using AEDs and CPR to resuscitate people who collapse outside a hospital setting: the survival rate for patients is low, only about 10 percent. If you do survive, you will probably have so much brain damage that you will never be able to resume your previous life. Will flying AEDs change those bad survival odds? Maybe, but maybe not. Drones might get there sooner than an ambulance and bystanders might be able to use the AEDs more easily, but the time window for saving a life with an AED is still only five to six minutes. Even with a system of centrally located autonomous flying drones, AED might not be started soon enough to make a difference.

A lot of resources and money are going into healthcare drones, even though not every idea will work out in the end. I feel strongly that this is still a good thing. If you aren't prepared to waste money on things that might not work out, you cannot possibly do things that are transformative. For every transformative idea, there are five times as many unsuccessful ones. That's why so much capital is going into this area. No one can figure out in advance which concepts are going to succeed and which ones aren't. It's easy to raise objections, but you have to try all the new ideas anyway.

A good example is the way drones were used, with FAA permission, to deliver medication to patients at a clinic held in Wise County, Virginia in July 2015. Wise County is in mountainous western Virginia. The annual clinic serves 3,000 patients over the course of a single weekend. The drones that delivered medicine to the clinic were the first ever in the US to have official permission. There were problems, mostly because small autonomous drones just can't carry big payloads, but a number of patients were able to get medicine they needed immediately without having to wait or travel long distances. As a proof of concept experiment, this one was brilliant.

Ambulance drones have lots of potential uses. A large AI-enabled drone could carry an emergency healthcare provider directly to the scene of an accident. The patient could be stabilized and sent on to the emergency room in the same drone, taking the place of the emergency provider. This is a big step up from a rescue helicopter. It's much less expensive, and because it's AI-enabled, it doesn't need a pilot. Because drones are much smaller than helicopters, they can land safely in places a helicopter can't. A drone avoids the traffic that delays ambulances.

Drones make great delivery vehicles for medication and supplies in remote areas or places that have been cut off by bad weather or natural disaster. A company called Matternet, cofounded by Andreas Raptopoulos and Paola Santana, got its start at Singularity University back when I was a VP there. Matternet has designed a drone exclusively for transporting things. The drone is an electric quadcopter with GPS and sensors. It can carry a kilogram of drugs, vaccines, or other supplies over 20 kilometers on a single battery charge. To extend its range, the drone can land at a network of solar powered charging stations and change its own battery. The operating system is simple and intuitive—easy to learn. The drones fly along routes that are the straightest line possible, while avoiding hazards and restricted air space. They even have parachutes in case something goes wrong.

In many parts of the world, healthcare is days of travel away over roads that are often impassable some of the year. Even in the US, many rural areas are medically underserved and healthcare facilities are hours away. Drones can deliver a needed medication quickly. Matternet can leapfrog over the missing road infrastructure in these countries by building a futuristic Pony Express that uses autonomous quadcopters instead of horses and riders to deliver supplies. Matternet is running prototype operations in some very remote places like Bhutan in the Himalayas, where there is only one doctor for every 10,000 patients.

In the bigger picture, drone systems like Matternet solve the last mile problem. How do you get something that last mile from where it is to where it has to be in a hurry? Flying drones are the most efficient approach. Jeff Bezos of Amazon predicts that seeing delivery drones will one day be as common as seeing mail trucks. He may be right. In 2015, Amazon received FAA permission to test its delivery octocoptors in the Seattle area.

MOVING ROBOTS

Moving robots are poised to enter healthcare, especially in the home. It's likely that many of them will be based on the PR2 personal robotics research and development platform. This open-source robotic platform, developed by Scott Hassan's company Willow Garage, has already been extensively used to create moving robots that do useful things autonomously. The PR2 platform has been used to create a robot that can fold a towel. That might not sound like a big deal, but for robots, it is. Folding a towel requires a lot of robotic dexterity.

The home healthcare robot has huge potential. This would be a robot that would assist you around the house. It would be a combination concierge, nurse, butler, companion—whatever you want or need it to be. It will have an AI component, so it will be able to move around your house easily. One thing it could do would be to pick you up out of bed and lift you places. In fact, robots that move you are already in use in some hospitals. A robot that could do this for you at home would be wonderful for people with mobility problems. It could restore a lot of independence to them. It would also be great for nurses, home health aides, and other caretakers. Lifting patients is a major source of shoulder and back injuries for these people and can sometimes also injure the patient. The robot could also bring you things, help you dress, remind you to take your medicine, and just talk to you. In hospitals, robots could complement humans by helping with patient care, bringing medications, and providing companionship.

Does all this sound a little creepy? Would seeing a robot instead of a person be disturbing to a hospital patient? Well, maybe— at first. We'd get used to them very quickly. How long did it take you to get used to carrying around a 2 x 5 computer in your pocket and talking to Siri on your iPhone? Or listening to directions from your GPS? We've already seen a lot of acceptance and even emotional response to household robots. From the late 1990s until 2005, Sony made the AIBO (Artificial Intelligence Robot), a robotic dog designed to be a pet. People became extremely attached to their doggy robots and were very upset when Sony discontinued them. Videos showing someone kicking a robotic animal really upset people, even though they know intellectually it's not any different than kicking a car or a soccer ball. It won't take us long at all to accept and even become fond of, or fall in love with, our AI nurse, doctor, and caretakers.

Once you adopt the technology, it's hard to remember what life was like before that. Think about life without your smart phone. We'll feel the same way about AI robots with great manual dexterity once they're in use and we've started interacting with them. We'll accept them pretty fast. We've already been working with robots for years and years and years in automated situations like factories. We've already automated a lot of things in our lives to the point where they're very routine. We're already working with them. Moving, interacting robots are just taking it to another level.

TELEPRESENCE ROBOTS

InTouch Health actually has robots in the hospital already. They're FDA approved, and they can go up and down the hallways by themselves. They're telepresence robots that are a way to improve telemedicine by bringing the camera and screen to the patient. The robots are very accurate; they can shine a light into a patient's eyes and detect if the pupils dilate.

Remote surgical collaboration, or telementoring, offers many opportunities to advance training and support of new and complex procedures and techniques. The InTouch Vantage system is a surgical collaboration platform for use in ORs and procedure rooms. The remote doctor can easily see everything by using remotely controlled pan-tilt-zoom cameras and a nine-foot articulating boom. It's great for remote surgical collaboration on an actual operation, and it also is great for training people in new procedures and techniques.

For bedside care, the FDA-approved InTouch Lite is already being widely used in clinical settings. This device doesn't look like a humanoid robot; it's more like a rolling platform with a large video screen. In fact, it looks a lot like a rolling blood pressure machine combined with an IV stand. It's primarily used for patient monitoring in situations where something might need quick action from a human. The device has a built-in stethoscope and ports that let other devices and monitors be attached. Imagine that you're in a rural hospital area far from a specialist in your particular healthcare issue. To see a specialist would be a long trip that you might be too sick to make. With InTouch and similar telepresence robots, the doctor comes to your hospital room and consults with you and the staff. If you need a procedure, the remote doctor can watch and instruct whoever does it. You get the expert care you need on the spot.

RP-VITA (the RP stands for remote presence) is an autonomous robot made jointly by iRobot (the company that makes the Roomba vacuum) and InTouch Health. These two companies combined have over 30 years of experience developing robotics and telemedicine devices. The RP-VITA is FDA approved for managing hospitalized patients in an acute care setting, meaning the patients are pretty sick and might need quick action. The robot is also approved for monitoring surgical patients before, during, and after their operations.

Scott Hassan is the founder of Suitable Technologies, which

makes Beam telepresence robots. Scott is a visionary in robotics. He started Willow Garage, which developed the PR2 robotics platform I mentioned earlier. His passion is bringing robotics into everyday life. That's exactly what Beam devices do. The BeamPro telepresence robot lets you interact with people in remote locations by coupling high-end video and audio with the freedom to move round. In simplified terms, it's sort of like walking around with a tablet computer and talking to people using a video link. This is an amazing concept for healthcare. It can give disabled people a way to work or go to school from home. It allows telemedicine consultation, which is great for busy doctors, patients, and training healthcare workers. It can also help elderly people stay in touch with family, friends, and care providers.

The rolling Beam robots are about five feet tall with a very streamlined look. The speakers and screens are on top. There's no attempt to make them look human. They're simply tools that can move alongside you or that can be sent to visit someone else. In 2015, President Obama used a Beam Pro telepresence robot to speak with Alice Wong, founder of the Disability Visibility Project, at a White House reception. I've had meetings myself using telepresence robots. I was once walking around at the XPRIZE Foundation office and had a meeting with someone who lived an hour away and was controlling the robot from home. She followed me down the hallway by telepresence robot; we talked face-to-face at eye level. It was amazing. I quickly forgot that she was there only on a screen and not in person.

Want to buy a Beam robot? You can do it at the fully robotized store on University Avenue in Palo Alto, down the street from my home. All the store clerks are BeamPro telepresence robots. There are no humans in the store. When you go to buy a telepresence robot on wheels, you get sold one by a robot!

Fellow Robotics, founded by Dr. Dan Barry and Marco Mascorro, is producing telepresence robots that are designed to be in retail stores. They're currently being used in Lowe's hardware stores. These robots are real game changers. You can go to conferences with them. You can go to meetings. They bring remote workers much closer to their office colleagues. In healthcare, you can have a medical conference where everyone involved in the care of the patient gets together to discuss the case, including the patient and the patient's family. Unless you've tried to do it, you have no idea how hard it is to organize that without some help from technology.

Chapter Seven

3-D PRINTING

AN IMPORTANT TREND NOW IS THE MAKER MOVEMENT, CUR-
rently intercepting with patients. Patients hacking devices for
their diseases with simple modifications. Patients are making
a difference by driving solutions in their own care on an individual case by case basis. This is going to get really exciting fast!
Widespread, easy access and knowledge about 3-D printing is
going to take the patient maker movement to a whole new level.
TechShop is leading the way in the maker movement, and Mark
Hatch (CEO) is fostering the intersection of makers and healthcare in a big way. I expect to see rapid movement in the area of
patients who suddenly have easy access to new tools, creating
solutions that can make major impacts, especially in daily life
tasks in their own lives. Traditional medicine doesn't address
the individual at this level of personalization, but patients can
and are, sharing with each other on patient communities.

To me, 3-D printing, also called additive manufacturing, is one of the most disruptive technologies out there across every single industry and economy. It gets especially exciting when you talk about it regarding healthcare.

The biggest concept behind 3-D printing is "Complexity is Free." What does that mean? It means that it will cost the exact same amount of money to print a 3-D medical device or brace for me that's perfectly fitted to my body as it does to print out a perfectly fitted brace for anyone else's body. The Invisalign clear braces people wear to straighten their teeth are created by a 3-D printer making a model of the patient's teeth then molding the plastic trays to the model. They are perfectly fitted to the patient and the patient's treatment plan and easily produced at mass scale.

The explosive growth of 3-D printing means that you can personalize and individualize literally every single thing in your life to your exact specifications—color, design, size—and it won't be more expensive. This is a huge difference from making one-size-fits-all! Anything that goes on or in the body will be printed to fit you 100 percent perfectly. Just about anything that was previously made of plastic or metal can now easily be printed in 3-D.

The impact of 3-D printing on medicine will be vast. It will affect everything. Just to take one example, external wearable devices like casts and braces will be made by 3-D printers and made from plastic, which can be exposed to water. Old style casts are clunky and uncomfortable; they're itchy, they're bulky, and they don't fit comfortably under clothing. You can't get them wet, and after a while, they get raggedy looking and smelly.

A 3-D cast is thinner and lighter than a traditional cast. It would be waterproof and could have openings in it to make it easier to wash and scratch. Finally, a cast you can shower with! It could be made with colors and designs so it would be more fun to wear. The best part is a personalized 3-D cast won't cost any more, and might even cost less, than a traditional cast. The cost to add in some personalized design is basically zero.

We're already seeing some really amazing medical uses of 3-D printing. In 2015, a patient with chest wall sarcoma had surgery to remove the tumor, which meant removing his sternum and some of his ribs. In a first for medicine, the missing bones were replaced with a titanium implant created from high-resolution CT scans that were used to program a specialized 3-D printer. Twelve days later, the patient went home and is doing well.

Another amazing use for 3-D printing is making drugs personalized precisely for the patient. In 2015, the FDA approved the first 3-D printed pill, a drug that treats epilepsy. For the best results with this drug, the dose needs to individualized for each patient. That meant using pill splitters to cut up the standardized pills and get the right dose, or close to it. The FDA approval lets the manufacturer use a 3-D printer to make pills in exactly the right dose for each patient. This approval will open a floodgate of 3-D drugs for many other conditions, such as Parkinson's disease, where getting the dose exactly right really makes a difference to the response.

Right now, 3-D printing in plastic is extremely easy for anyone to do at home. You can buy a printer that runs from your home computer for under $500. The more complex ones run in the thousands and tens of thousands; the most expensive 3-D printer can run upwards of $1 million. To make something with a 3-D printer, you start by creating digital files on your computer using computer-aided design (CAD) software. The files tell the printer how to build the three-dimensional object by building it up in layers. They're sort of like inkjet printers that work in three dimensions.

Currently you can print using one type of plastic with an inexpensive home 3-D printer, but you can do a lot just with that. Add in printers that work with several different types of materials at once, including metals, and the possibilities multiply exponentially. At Stratasys, one of the biggest manufacturers of 3-D printing hardware, the technology can now make amazing 3-D models of body parts so that surgeons can visualize a complex brain operation in advance. The printers use biocompatible materials to make custom devices that will be implanted in the body and custom designed exoskeletons. Ekso Bionics, mentioned in a previous chapter, has already combined their exoskeleton with 3-D printed pieces created by Scott Summit, a pioneer and thought leader in the 3-D printing space. Scott was the founder and CEO of Bespoke Innovations, which specializes in completely personalized prosthetic limbs—an area where 3-D printing has already become a standard technique.

Right now, 3-D printing is mostly being used in major medical research centers. As the costs and accessibility for advanced 3-D printers goes down, we'll see them in every hospital, emergency room, and doctor's office, printing personalized versions of whatever is needed. It will be 3-D printing on demand, instead of 3-D printing from a lab or factory. For personal use, we're already seeing 3-D printers at the local copy and print shops. There's a start-up company experimenting with 3-D printing vending machines. Email or upload the design you want printed to the website, have it printed out overnight, then pick it up from your private drawer in the vending machine the next day, or even a few hours later, when it's ready.

Of course, for implanted medical devices, 3-D printing at home or a copy shop won't work, at least not yet. The medical-grade objects that are going to be implanted in your body need a much more sophisticated system that can use multiple materials. They will also need to be FDA approved. These devices will have to be very precisely created in a special lab at the hospital and completely sanitized. The amazing part is that they will be completely and utterly personalized for you and you alone. Until recently, if you had to have a trachea implant, it was going to be one made in a standardized size. Now, we have the technology to make it in exactly your size.

Where once a part or device had to be in stock or ordered from the manufacturer, now it's made on demand on the spot. What does that mean? Among other things, no costly inventory tying up money and taking up space. That could lead to cost reductions. 3-D printing will be very disruptive. When we all have 3-D printers in our houses, printing whatever we want, every day, on demand, how will that change shopping and manufacturing? How many businesses and entire economies are going to be dramatically altered or bankrupt? In a world of everything made on demand, does everything become disposable and recyclable? At the end of the day, would you just throw everything you made and used that day into a 3-D printer recycler to break down the materials and reuse them? Imagine a world where everything is personalized, made only as needed, used as often as needed, and then recycled,while costing less than the same mass-produced objects cost today. If you imagine the implications across your entire life, this is going to disrupt every single thing you do. It's going to make things easier, faster, cheaper, better, more accessible, and completely personalized.

3-D PRINTING IN USE

Assistive devices made with 3-D printing and advanced materials will revolutionize the lives of people with chronic diseases and handicaps. I've already talked a bit about how exoskeletons will change lives. Where will the exoskeleton come from? A 3-D printer. Body scans will be used to design it to your exact body specifications, and then it will simply be printed out in pieces and assembled quickly.

A great example of 3-D printing for an external assistive device is the "arms" being made for young patients at the Nemours/Alfred I. duPont Hospital for Children in Wilmington, Delaware, using a standard Stratasys printer. In one case, a young girl with very limited arm motion was given a custom-built assistive device that is basically an exoskeleton, an assistive device made of hinges and resistance bands. It straps on like a vest; the "arms" simply run along the child's arms, over any clothing. The exoskeleton lets kids with underdeveloped arms use them for normal functions, like feeding themselves, brushing their teeth, and hugging. Because 3-D printing uses lightweight plastic, the exoskeleton is light enough for even very young children to wear, letting them get started on physical therapy much younger and potentially avoiding developmental delays. As the child grows, larger, perfectly fitted devices can be created very quickly. A dozen or so of these exoskeletons are in use today. The videos of these kids leading lives that are now so much more independent are amazing.

One of the most fun aspects of 3-D printing to me is that it lets us turn handicaps into personal fashion statements. We can be creative about prosthetics. Instead of trying to hide a prosthetic hand by making it look as close to a real hand with flesh tones as possible, why not make it bright red with yellow stripes? It doesn't have to mimic how a hand looks; it only has to mimic how a hand works, and even that can be improved upon dramatically! Research is moving ahead very quickly on prosthetic limbs that are controlled by the brain and move in very natural ways. For example, hands that grip firmly and let each finger move independently may not be far away.

Why try to hide something when you can have fun with it instead? People we would consider severely disabled today will soon be super-enabled. Take Aimee Mullins, a double leg amputee who is equally well-known as an athlete, a fashion model, and a spokesperson for people with disabilities. She has several sets of prosthetic legs. Depending on which legs she chooses to wear, she can change her height by several inches.

She has printed legs, decorative legs, and legs that enable her to run faster and jump higher than an average person. Would you call her disabled? I don't. She took a devastating situation and turned it around to become super-enabled and excel.

The big deal with 3-D printing is that you can do anything with it. You can reimagine the human body and improve on it. You can do things that are incredibly artistic and you can do things that are very functional and make massive improvements on the human physique. A fashion statement, a new type of functionality, or both.

Think about it. You need a replacement arm. You can now use 3-D printing with multiple materials to create the mechanical part of the arm in a way that fits you perfectly. Your new arm is going to be so comfortable and effective that you'll want to do the physical therapy necessary to learn how to use it. You'll wear it more often and be able to do more for yourself. But your prosthetic arm is actually better than a real arm because you can make it more functional. You can make it rotate 360 degrees, for instance, or be longer than your natural arm. By taking this opportunity to make improvements, you can become not just enabled but super-enabled. And then you can design the external part of the arm to look any way you want.

COMFORT AND COMPLIANCE

Compliance and adherence are big problems in medical treatment. We don't always take our medicine on schedule or even at all. We don't always use our medical devices—a knee brace, for instance—when and how we should. 3-D printing will really improve compliance with some medical devices by making them much more comfortable. Let's take scoliosis, or abnormal curvature of the spine, as an example. The majority of patients with scoliosis are girls in the 9-15 year old age range. The treatment for scoliosis is to wear an ugly, clunky, uncomfortable back brace for as many hours a day as possible, literally for years. The more you wear the brace, the better off you'll be, but what teenaged girl wants to wear something so horrible? Not surprisingly, compliance and adherence are big problems for treating scoliosis.

With 3-D printing, there's an opportunity to dramatically change that—to make wearing your back brace more comfortable and even fun. A back brace made with 3-D printing will be based on body scans of the individual patient, so it will fit exactly right with no gaping or rubbing. It will be perfectly aligned to exactly what your shape is. If desired, it will be made to be worn on the outside over clothing. You'll be able to choose the colors and patterns yourself, or maybe incorporate designs from a famous artist. Instead of being an ugly medical device, it will be a massive fashion statement. Now you'll wear it because it's pretty and it's cool. It's also a lot more comfortable because it's lighter, more flexible, easier to get on and off, and fits a lot better.

Here's where we can start to see a convergence with other technologies. Wellinks, a Kairos Society company, is incorporating sensors into back braces that track how long you wear the brace each day. That tracking will send information to your healthcare provider and maybe your parents, but it can also be used to incentivize you to wear your brace. Potentially, the information goes to iTunes as well. For every day you wear the brace at least twelve hours, you could get a free download. The added benefit of feedback loops also ups compliance, as when patients know they're being accurately watched and cared about, they'll behave differently and compliance goes up.

INTERNAL USES

For internal implants, look at the recent cases of babies born with a problem in their tracheas (windpipes) and bronchial tubes. They were missing some of the supportive cartilage that keeps the tubes open. These kids had episodes of not being able to breathe because their trachea or bronchi had collapsed. They were always at risk of sudden death.

No surgical solutions were good options for propping open the airways. But with 3-D printing, surgeons were able to use CT scans to make exact replicas of the damaged airways. These were then sewn around the actual airways to wrap them in a sort of sleeve and provide the missing support to keep them open. Without 3-D printing, this would have been impossible. With it, these kids will lead normal lives. The supports are made of a biodegradable material that will be gradually absorbed by the body over several years. By then, the airways will have grown enough so that the support is no longer needed.

Even more amazing is that we can now print a completely artificial, custom-made trachea out of silicon and transplant it into a patient. It's already been done successfully several times, including for a little girl born without a trachea. And plastic scaffolding made using a 3-D printer has been used as the support for growing an entirely new trachea using the patient's own stem cells.

The same underlying scaffolding idea is being explored for growing new teeth right inside the mouth. The scaffold would be personalized for you and created using a 3-D printer that exactly duplicates the missing tooth. The biodegradable scaffold is implanted and populated with your own stem cells. Once the stem cells colonize the scaffold, the tooth can grow in the socket and merge with the surrounding tissue. The tooth regenerates in a few months.

If you need a medical device, temporarily or even permanently, why should you be stuck with something ugly? Why not completely personalize it into something that is representative of you? People get tattoos all the time. They do piercings, they change their hair styles, they dye their hair, they put on make-up, and they wear different eyeglasses. They use fashion as a way of presenting themselves to the world. Why not do that with everything else on your body? Why not personalize that and make it an extension of your artistic nature? Or take it a step further. Make yourself and what you're wearing more beautiful and interesting while wearing the device than you would be without it.

What's so amazing about 3-D printing is just how empowering it is for people with handicaps and chronic diseases. You'll be able to print a lot of what you need without having to ask other people to get it for you or struggle to get it yourself. That may seem very far-fetched today, but let's go back to the VCR analogy. When VCRs first came out in the 1980s, they were expensive and hard to figure out. Very quickly, however, everyone had one. When DVD players came along, they quickly replaced VCRs because they were cheaper and better. Now we have TiVo and streaming. Every new technology that has come along has changed how we watch TV. Each one has been widely adopted faster than the last. The same thing is happening now with 3-D printing. As the printers cost less, get easier to use, and do more, they'll be more widely adopted, and we'll discover new things to do with them. In twenty years, you might be able to print your own clothing. You'll be able to use a 3-D printer to print a 3-D printer. Doesn't *that* blow your mind? Inanimate objects will essentially reproduce asexually. Instead of rapid iteration, we can have rapid evolution of baby printers.

For healthcare, 3-D printing is going to do a lot to reduce some costs, like for prosthetics. Today, an artificial limb is very expensive. With 3-D printing, costs will go down and access will go up. The devices will fit better and be more comfortable, so people will wear them more and be more independent. 3-D printing will improve compliance in other ways as well by making any medical device fit better and be more comfortable.

In some conflict areas of the world, many people have had limbs destroyed by bombs and land mines. Even if they could afford artificial replacements, they don't have access to skilled professionals to make them. With 3-D printing, we could really help these people. A great example is the organization e-NABLE, which describes itself as a global network of passionate volunteers using 3-D printing to give the world a "helping hand."

Volunteers create custom artificial hands at low cost or for free for anyone who wants one, using designs they download from the organization's Hand-O-Matic software or designs they create themselves using e-NABLE's basic Raptor Hand design.

One of the problems with printed prosthetic hands is that the plastic used to make them, especially one called PLA, melts if it comes into contact with anything too hot. To solve this problem, designers are working on putting temperature sensors in the hands. Plastic hands are controlled by the wearer's muscle movements. By placing temperature sensors connected to a microcontroller in the hand, a signal is sent to the muscles telling them to ungrab the hot object.

Imagine a one-room clinic in a remote area of Africa. Getting medical equipment to it is costly and time-consuming. What if the clinic had a solar-powered 3-D printer instead and could simply produce things like syringes and scalpels as needed on the spot? The problem of keeping the clinic supplied with basic medical equipment becomes a lot simpler. Even in the US, it can take a few days for specialized medical equipment to arrive at a rural hospital. 3-D printing would remove delays and shipping expenses.

What works for a rural clinic also works for a large modern hospital. Right now, hospitals tie up a lot of storage space and money to keep on hand medical supplies that aren't needed all that often. Hospitals could use 3-D printers to make these items as needed instead.

And what about future medical needs for long space voyages or to colonize other planets? Made In Space, cofounded by Jason Dunn, makes very sophisticated 3-D printers that work in zero gravity. One is in use now on the International Space Station to make small parts. This is a lot faster and more efficient, to say nothing of cheaper, than sending parts up by supply rocket. Made In Space got started with a $10 million grant from NASA, but it's now funded by investors.

Many people are surprised to learn that 3-D printing has actually been around for decades. That means some of the basic techniques are now coming off patent and becoming much more widely available. We're going to start to see massive innovation in the field as people expand on the existing technology. We already have advanced printers that can combine ten different materials when making something. We'll see 3-D printers become as ubiquitous and disruptive as smart phones. The world is poised and waiting for it.

BIOLOGICAL 3-D PRINTING

The next frontier is biological 3-D printing, also known as tissue engineering. It's the combination of cells and 3-D printing. One good example is what a company called Organovo does. Gabor Forgacs, one of the pioneers in biological printing, is the scientific founder of Organovo and has spent a lot of time talking with me about his outlook in this area. One of their products is a 3-D liver assay that can be used to test drug toxicity. A tiny patch of human liver cells are grown on a 3-D scaffold inside a bioreactor. These cells react to drugs much as a liver inside you would. This is a huge step forward for drug development. No longer do drugs have to be extensively tested on living animals and humans to see if they might be toxic to the liver. They can be quickly tested using a liver assay instead. The next step from that would be to do something similar with an individual's liver cells. That way, if you need a drug that might cause liver damage, you can tell in advance if that will happen to you. Right now, all you can do is take the drug and hope for the best on a liver function blood test later on. Organovo is also working on kidney assays for the same purpose.

Big Pharma is very interested in what Organovo and other companies are doing with bioengineered tissues because it could hugely disrupt the way drug trials are done. At present, it takes on average between $1 billion and $4 billion and ten years of studies to bring a new drug to market. What that means is pharma companies concentrate on very expensive drugs that are useful only to a small subset of the population. If 3-D bio-engineering can make the clinical trials process much, much faster, it will save vast amounts of money and make valuable drugs available more quickly. We'll be able to learn right away what side effects and toxicities an experimental drug might have, without testing it on animals and humans first.

A really exciting area for biological 3-D printing is in skin grafts. This will have a massive impact on patients with burns and hard-to-heal wounds or have had radiation treatments. For skin grafts today, skin is taken from another part of the body and grafted to the damaged area. It's a slow and painful process. A lot of research on skin grafting has focused on ways to use less skin or make it grow in faster. With tissue engineering, the paradigm for skin grafts shifts. Scans are used to determine exactly where the skin needs to go and at what depth, which is important because different layers of skin have different types of cells. The patient's own stem cells are used to grow the right sorts of skin cells in a hydrogel. One of my heros, Dr. Bob Hariri, has done extensive work using stem cells to treat burn victims at Celgene Therapeutics. A 3-D printer lays down the hydrogel in the damaged area just as an ink-jet printer lays down ink. The cells grow new skin exactly where it's needed. The end result is faster healing with much less scarring and less contracture of the skin and underlying muscles. Right now, we have printers that can lay down two layers of skin, deep enough to treat and to heal most burn wounds. Clinical trials could start within the next five years, pending FDA approval.

Biopen is a product from a company out of Australia, a pen surgeons can use to lay down stem cells onto an injury by using the patient's own fat cells (good news!). The cells are 3-D printed inside a biopolymer and has a second layer of gel material for protection. Then the Biopen lays down layer after layer of stem cells to treat the injured area.

Tissue engineering to build organs could help reduce or even eliminate the need for transplanted organs. We already have a shortage of transplant organs that will only get worse in the future. Right now, many organs come from people who die in car and motorcycle accidents. In the future, self-driving cars will mean far fewer fatal accidents, which in turn means even fewer organs for transplant. I'm personally working in this area of transplants with the Department of Defense, OSTP, and DARPA to help make massive amounts of funding available to labs that are doing organ preservation and tissue engineering.

With fewer organs available, it becomes even more urgent to learn how to preserve and grow new organs instead.

We're making some progress here. In the lab, it's now possible to grow hollow organs such as bladders and some types of cartilage, such as ears and noses. We can grow these and things like the liver assay cells I mentioned earlier because they don't need a dense network of blood vessels to survive. Vascular organs such as hearts, kidneys, and livers are much more complex. They need to have oxygen and nutrients delivered to every cell. One possible way to do this is to take an existing organ such as a liver and decellularize it—remove all the cells but leave behind the supporting structures and blood vessels as a scaffold. Stem cells could be used to populate the scaffold and grow a new liver. We can do that now with the bladder because it's basically hollow, but bigger, more complex organs will take more work.

This is another area where 3-D bioprinting has huge potential for breakthrough technology. At the Wyss Institute for Biologically Inspired Engineering at Harvard University, Jennifer Lewis leads a team that has figured out how to print blood vessels in tissue using disappearing ink.

The tissue is built up in layers with a 3-D bioprinter. The "inks" are viscous gelatins that contain structural supports and living cells. To make the blood vessels, a different type of gelatin is used, one that is viscous at room temperature but liquid when it's cooled. This ink is used to print blood vessels within the cell layers. When the patch of tissue is cooled, the "invisible ink" of the blood vessels is gently sucked out, leaving the channels behind. Epithelial stem cells, the kind that make the linings of blood vessels, are seeded into the channels, where they will grow into working capillaries.

OK, this is pretty wild stuff, and it's not even close to clinical trials yet. In fact, it could be fifteen, even twenty-five, years before we can grow functioning replacement organs. But the amazing thing is that the research is happening. This is an industry in its infancy. We're just getting started with the technology, but the potential is very clear. Organ failure is the number one killer in the US, and about 35 percent of deaths in the US—900,000 people a year—could be significantly delayed or prevented through an organ transplant. Imagine a world where instead of repairing an organ or treating a disease with medication, you just get a new one grown from your own stem cells.

Another very exciting area for tissue engineering is growing replacement bones. Right now, badly damaged bones that can't heal themselves are treated using titanium implants or bone grafts taken from elsewhere in the body. They work, usually, but not as well as your own natural bone would. EpiBone, a company founded by biomedical engineer Nina Tandon, uses 3-D printing to create bone grafts using the patient's own stem cells. The process starts by using CT scans of the damaged area to make a very precise 3-D model of it. Stem cells are then taken from the patient. The 3-D image is used to design the graft and a custom-built bioreactor for it. The stem cells are seeded into the graft, and it's placed in the bioreactor. The stem cells turn themselves into bone that can then be grafted back into the patient. Because the graft is from the patient's own cells, the body won't reject it. In fact, the body integrates the graft, and it becomes living bone tissue. Human trials for the EpiBone procedure are only a few years away.

Another approach to bone grafting uses 3-D printing with a biocompatible plastic called PEKK (polyetherketoneketone). It's similar to bone in density and strength, and it can withstand the high temperatures needed for sterilization. PEKK is being used now with 3-D printing to make personalized facial and skull implants to replace damaged and missing bone. The implant is printed with tiny perforations to allow the patient's own bone cells to grow in and attach to it. Because it's plastic, the patient can have MRI and CT scans, and it shows up on X-rays. In 2014, the top of an entire skull was printed using PEKK and implanted in a woman's head. The operation was successful. The surgeons were able to literally watch for blood clots and swelling in the brain after the surgery.

So much of what I've talked about in this chapter has been talked about for years as something that might happen sometime in the faraway future. Now, it's coming into focus. The potential for radical change is vast and very, very exciting.

Chapter Eight

THE FUTURE OF
THE HOSPITAL

A LOT OF WHAT HAPPENS IN A HOSPITAL TODAY IS GOING to happen in the future in the patient's home or at a nearby pharmacy or storefront clinic. All the point of care diagnostics, the sensors, the wearable technology, the cloud-based computing, the telepresent robots, and technology we haven't even thought of yet will converge to make hospitals places where you will go a lot less often. Even people with chronic diseases and severe handicaps will be able to get excellent medical care without leaving home. And people who live far from advanced medicine will still be able to access it.

When I was growing up in New Hampshire, our small town didn't have a hospital. The closest hospital was several towns over. Specialists and more advanced hospital care were a long drive away. I sometimes forget how lucky I am now to live in

California and be surrounded by high-end hospital systems like Stanford and UCSF and PAMF. I didn't have that access when I was younger. Many people still don't. If you're in rural Wyoming, you might need to drive for hours to see a doctor. And if you live in rural parts of Africa or India or anyplace where the nearest doctor could be a two- or three-day walk away, your access to high-level medical care is basically nonexistent. The medicine of the future will give the world access to healthcare.

Telepresent medicine in the US is starting to take off after being around in various forms for decades. The old-fashioned house call is back in the form of high-tech in person visits. A number of companies now offer ways to connect with a doctor remotely for a "house call" at any time of day or night. As many medical practices are discovering, this is a great way to offer patients extended service hours without having to actually keep the office open.

Telepresence medicine can mean a lot of cost savings for patients and insurers. A handicapped patient who might have needed costly special transport to see a doctor can now do it remotely instead. Sensors that provide information such as pulse and blood pressure can automatically transmit that information to the doctor or healthcare provider, who also has all the patient's electronic medical records. The two-way discussion of symptoms and treatments can happen very much as if it were in person.

A recent survey of 444 midsize to large US companies that employed about 7.2 million people in all, found that almost two out of every five employers now offer a virtual doctor's visit as a healthcare benefit. About 20 to 22 percent of employees in the surveyed companies were taking advantage of telepresence medicine benefits just in 2014 alone. By 2018, that number is expected to hit 81 percent. This is definitely an industry on a massive upswing. You can do telepresent medicine through Skype or Facetime or with telepresent robots that let doctors beam in to your bedside. In fact, some telepresent robots are FDA approved for hospital use.

In the hospital, being able to speak directly with the doctor using a telepresent robot makes being a patient more efficient. When a patient develops a new symptom, for instance, the patient or maybe a family member tells the nurse, who calls the doctor and discusses the situation, or more likely leaves a message and has to wait for a call back. When the call comes, the nurse has to relay the information to the patient. All the back-and-forth could take an hour or even longer. Instead, the doctor can beam in to the patient's bedside and talk to you, your family, and your nurse directly. It's faster, and there's much less chance of important information getting dropped or garbled along the way. The patient gets the right care sooner.

As a hospital patient, I feel much more secure if I can see my doctors, even if it's through a telepresent robot, when something comes up. Otherwise, I know I won't see them until they stop by for a few minutes during morning rounds the next day. Telepresence lets doctors see many more hospitalized patients over the course of the day than if they were on foot going in and out of patient rooms, looking for patients who aren't in their beds, and getting stuck in hallways with family members and staff who need questions answered. Telepresent robots maximize a doctor's time and give more attention to patients. With the robots, now I'm getting first-hand care versus care relayed to a nurse through a paging system.

Doctors aren't the only ones who will be telepresent. A company in San Francisco called Sense.ly provides virtual nurses using AI and avatar-based technology, incorporating Microsoft's Kinect motion tracking software and Nuance speech recognition software. Nurse Molly has an excellent, empathetic bedside manner. Because Molly uses AI, she's a smart nurse. She does more than just provide answers to questions or triage your symptoms. She provides virtual treatment.

She's great at monitoring your symptoms doing follow-up care if you have a chronic disease. Even better, she can do it in English, Spanish, Mandarin, and 24 other languages. This is a wonderful tool for patients who don't speak English well or at all. Things like medication compliance really go up when your drugs are explained to you in a language you understand well. No longer will family members, even kids, be pressed into service to interpret medical terminology they don't understand. The information the patient needs will be presented directly. It works the other way around as well. Let's say I get sick while visiting France. I speak French, but not well enough to negotiate my way through a French hospital. Molly could come to my rescue.

Sense.ly could also save a lot of unnecessary trips to the ER or urgent care clinic. I can talk to Molly first and see what she thinks. If she says to go to the hospital and I don't speak English, she'll be there to speak to me in my own language. Sense.ly can be used anywhere, so it has big implications for medical practices to see more patients and provide more accessible care. Molly never sleeps, so you can talk to her anytime, even when the office is closed.

Telepresent medicine also now gives you access to the best doctors in the world, no matter where you are and no matter who you are. A company called Best Doctors provides exactly that access. You enter your medical condition and the algorithm can find you a doctor anywhere in the world who is a perfect fit for you. The appointment takes place on line. Best Doctors actually has 53,000 doctors signed up, including the top 5 percent of all US physicians.

LABS ON A CHIP

The lab is now coming out of the hospital as well. It's moving to the pharmacy and local clinic. This is great for patients because going to a hospital or stand-alone lab can be very time-consuming. If you have a health problem, just getting to the lab can be difficult and exhausting. If you're immunocompromised, the last place you want to be is in a hospital lab, sitting around in the waiting area sometimes for longer than an hour! You could easily catch something from another patient.

When I need blood work as I often do, instead of a long drive to a hospital lab and sitting around in a waiting room, I now pop in to a nearby drugstore and have my blood work done there on early mornings, evenings, and on weekends and holidays. I've never experienced more than a ten minute wait, and it's only two blocks from my house. That's already a major improvement, but in the future, the lab work itself is also very different. The amount of blood could be just a drop from the tip of the finger, using 1/1,000 of the amount of the blood needed for a standard blood draw done with a needle into a vein. Using microfluidics, that one drop of blood can be used to test for 30 different things using a disposable lab-on-a-chip. The results will be available almost instantly.

We'll start to see more and more labs get widely distributed in places like Walgreens, cvs, and other large retail chains. In most places, you currently need a lab order from a doctor to get lab tests done, but in the future, you'll be able to request at least some of your own testing. In the state of Arizona, you don't need a doctor's order to request an medical lab test. Self-initiated testing is a crazy new level of control for the patient. You could arrange for your own test for an std and keep it very private; if you have diabetes, you could have your A1C level checked to see how your blood sugar control is doing. If you really want to monitor your health closely, you could do tests for anything you wanted to chart on your own. As instant labs become more available, will Arizona's law spread to other states? Is it even a good idea? The debate between patient autonomy and healthcare provider oversight is ongoing.

Companies working on diagnostics using microfluidics are extremely innovative, but technical issues remain. When they're eventually worked out, lab testing will be not just faster and easier but also cheaper and in some ways better. Data from lab tests figure into about 70 percent of all medical decisions, so anything that makes them easier and cheaper for the patient will have a major impact on healthcare and healthcare costs. There's plenty of room for streamlining in the lab work process. Right now, the same lab test can have a wildly different price depending on where you go and what your health insurance will cover. The problem is that you the consumer have no idea what the cost actually is. When companies operate through massive chains like Walgreens and cvs, everything is transparent. You go on the website, and it will tell you exactly what each individual test costs; it's always the same, no matter where you're located. Transperency's a huge step forward in containing healthcare costs.

MACHINES DO IT BETTER

Blood tests still need a blood draw, and IVs require a needle into a vein. These needle pricks can be very difficult to do on some patients; you can't always find a good vein or hit it on the first try. Even an experienced phlebotomist hits the vein the first time only about 75 to 80 percent of the time. Robots can do it better. There is a company working on a robotic device for doing blood draws and starting IVs that, in prototype form, has an 83 percent success rate in hitting the vein the first time. It uses ultrasound and special imaging software to determine where the vein is. A human only has visual cues, but the robot has both sound and vision, making it more accurate.

Robotic phlebotomists are only prototypes now, but when this type of technology becomes standard, it will be a big time saver and pain saver from being pricked too many times. Just about every patient is going to need at least one IV started during a hospital stay and will probably need at least one blood draw. Nurses and technicians spend a lot of time on this. A robot that could take that over would free them up for other aspects of patient care. It would also do a lot for patient comfort and safety. There were times where I've had to have my vein stuck five times to be able to get the needle. Five times?! I've had almost 200 IVs in my life, and that's not even counting all the ones that they didn't hit the vein on the first time. When technology like this hits the market, and is at 93 or 95 percent efficacy in hitting the vein on the first try, I'm going to be the first to sign up.

Machines will also replace anesthesiologists for some surgery. Sedasys is an anesthesiology machine that has been approved by the FDA and is in test use in several hospitals. The idea is to replace one of the highest paid medical specialties during routine operations and free the anesthesiologists to be more available for complex cases that need their skills more. This could really improve the flow of patients through a surgical suite without compromising safety and efficacy.

Right now, Sedasys is mostly being used to provide anesthesia for colonoscopies, which are among the most common medical procedures—about 14 million are done each year. For patient comfort, they're usually done under anesthesia (fortunately!). In 2009, that meant $1.1 billion spent on traditional anesthesia services for colonoscopies alone. That increases the cost of a colonoscopy by up to about $2,000. Sedasys can provide the same anesthesia for $150 to $200. That's a massive reduction in cost. Sedasys is arguably safer than a human anesthesiologist. The machine monitors the patient continuously using sensors that keep the patient at exactly the right level of anesthesia. It never gets tired or distracted.

HOSPITALS AS HOTELS

Not every hospitalization can be planned in advance—emergencies happen—but a lot of surgery is scheduled ahead of time. Here's where a hospital will function more like a hotel. It will start with new designs for hospitals that are more welcoming and easier to navigate with more amenities for both patients and visitors. Hospitals are starting to take a page from the hospitality industry and are tying to make the customer experience more like staying in a hotel than in a hospital. Simple things, like being able to get a meal even after the kitchen and cafeteria are closed, make a hospital stay much more tolerable. Right now, some hospitals still serve patients three meals a day at 7 a.m., 11 a.m., and 5 p.m. If you want anything to eat beyond Jell-O and ginger ale outside those hours, you need to rely on a family member or friend to bring it to you. Some hospitals already let you order food on demand by asking a staff member. Soon, you'll be able to do it with an app and have the food delivered by a robot. Autonomous robots that deliver and remove food trays and bring you your medication and supplies will be common in future hospitals. Robots are efficient and accurate. They don't call in sick, they never get tired, and they never go on break. There's no human error from distraction or overwork. The robots free caregivers to do other things for the patient.

The real future of the hospital is being hospitalized at home. With the convergence of continuous monitoring and all sorts of sensors, wearable monitors, point of care diagnostics, and telepresent consulting, you can potentially get hospital-style treatment while staying home. This is actually happening now in some places. Mount Sinai Hospital in New York City has a home hospitalization program that has a nurse and doctor visit you every day. Lab draws and IV medications, even X-rays and ultrasound, can be done in your own home. In the Mount Sinai program, the cost for in-home hospital is no greater for patients than if they were physically in the hospital—with none of the annoying noises or dangerous infectious diseases you find in a hospital.

When you're well enough to be discharged, usually after three or four days, you're discharged to your own house. There's none of that horrendous waiting around in the hospital as all your discharge paperwork gets completed. Right now, in-home hospitalization uses physical visits from healthcare providers. In the future, robots in the home will provide the nursing care and the visits will be done with telepresence.

A hospital is a terrible place to be when you're sick. The worst part is you can't get any sleep. It's noisy all the time, even at night, the beds are uncomfortable, it's not your usual pillow, your roommate snores or coughs, you're in pain, maybe you're having side effects from medications, and you're separated from your family and pets. The stress of the experience is very exhausting and even dangerous to your health. It can slow healing, and there's always the chance of a hospital-acquired infection. The stress of being in the unfamiliar hospital environment can lead to older patients or patients with cognitive problems developing delirium and needing sedatives. That's not good for them.

Better patient outcomes with less chance of rehospitalization is the main reason companies and hospital systems are not just brainstorming but actually executing hospitalizations at home. Hospital systems are really under pressure to reduce costs and improve quality. An in-home hospitalization—essentially a house call to the extreme—is one great way to do it. All the accelerated technologies I've talked about throughout this book are enabling this to happen. Suddenly, you're going to be on the road to wellness from getting a good night's sleep in your own bed without all the horrible noises, beeps, and alarms that go on almost as much at night in a hospital as they do during the day. No one will come in at 6 a.m. to take your blood pressure, and no one will wake you with a breakfast tray at 7 a.m. You're home with your family. If you have pets, they're snuggling with you, which is extremely healing. Of course, mobile acute care, as some are now calling in-home hospitalization, isn't for everyone. You need to be relatively stable, have a safe home environment with caretakers, and be able to do most or all of your own self care.

Implementing a home-hospital program isn't easy. Mount Sinai had to create a complicated system of backups, because the patients need to have 24-hour access to a physician and nurse coverage. They had to make arrangements with emergency medical service providers. Patients who are hospitalized at home are followed for a month after their home-hospital stay. They're eligible for services like health coaching and home doctor visits to help prevent rehospitalization through better self-management. Even with all the complexities, so far the system is a win-win.

Over time, converging technologies and new approaches means that hospitals will become places only for serious emergencies, very sick patients, and specialized high-tech care and surgeries.

TELEPRESENT INNOVATION

Other innovative models for telepresent medicine are arising. Curely is the first mobile marketplace to bring together patients and doctors from anywhere in the world. All you need is a smart phone, which means doctors can easily supplement their physical practices by reaching about two billion people in the world. Curely is an app that brings board certified doctors from around the world to your smart phone without a middleman—a sort of Uber for medical care. You choose your doctor by browsing through the list by specialty, price, and language and checking the ratings and reviews. You can then chat live with the doctor or send an email message at anytime of day or night. The fees are very affordable, starting at $10 for a 12 minute chat session and $2 per email. You can approach the doctors directly without medical insurance or a referral. The Curely messaging platform uses IBM Watson; the live chat and email use a lot less bandwidth than most other telepresent applications. By keeping bandwidth down, Curely is targeting the 80 percent of the world's population that doesn't have 3G connectivity. Curely CEO Paul Lee estimates that the telehealth industry is going to grow to $4.5 billion by 2018. Mordor's Market Intelligence estimates that virtual care and telemedicine worldwide will be a $34 billion market by 2020, with the US accounting for 40 percent of that, nearing $15 billion.

In 2015, Mercy, a health system based in St. Louis, opened a $54 million virtual care center to house telepresent medicine programs. This is a very cutting edge operation. It's in a four-story, 120,000-square-foot building. Nearly 300 physicians, nurses, specialists, researchers, and support staff will deliver care 24/7 via audio, video, and data connections to locations across the Mercy system. Care will also go outside Mercy through partnerships with other healthcare providers and large employers. Mercy estimates that the center will manage more than three million telehealth visits in the next five years. Just as cool is the way the center also will be a hub for advancing telemedicine through research and training. This level of commitment by a major hospital system is going to give us huge insight into how to use telemedicine really well, both for patient care and for training and mentoring staff. In a few years, we'll also have some really actionable data on how telemedicine reduces costs in the long-term. It's much cheaper to operate a telepresent medicine service than a brick-and-mortar office. You can charge less for the same services and also save money by avoiding complications, ER visits, and hospitalizations. This is a win-win.

INNOVATIONS IN ACCESS

cvs pharmacies now have a service in some stores nationwide called Minute Clinic. Staffed by family nurse practitioners and physician assistants, Minute Clinics offer basic healthcare, such as administering vaccines, diagnosing, and treating minor ailments like earaches and urinary tract infections. They're open all day every day, including evenings and weekends—no appointment needed. Insurance is accepted, but cvs also posts a price list of exactly what each service costs. Transparency in medical costs is a new concept that some patient-facing companies are adopting, finally! cvs has 7,800 retail stores and a presence in almost every single state. Right now, it's the country's biggest operator of health clinics and the largest dispenser of prescriptions drugs. In 2014, it had $140 billion in revenue, 97 percent of it from prescription drugs and pharmacy services.

When you stop to think about it, a drug store chain like cvs can be the country's biggest healthcare company. It's already bigger than most drug makers and wholesalers. cvs is poised to really turn the healthcare industry on its ear. It's very convenient for patients to just stop in at their local cvs or Walgreens and be seen right away with no appointment. This level of easy access can make a huge difference in utilization. The Minute Clinic can help people with chronic problems manage them better by providing quick, easy monitoring. The nurse can help you avoid a visit to the urgent care clinic, your doctor, or even the ER by diagnosing and treating your minor problem on the spot. And the staff can also quickly realize that you might have a serious problem and send you on for more advanced diagnosis and treatment. At a typical cvs Minute Clinic staffed by nurse practitioners, they see 35 to 40 patients a day. The average bill to treat minor injuries and illnesses is $79 to $99. Most major insurance plans are accepted. The RAND Corporation has done studies estimating that more than a quarter of emergency room visits could actually be handled by retail clinics and urgent care centers. The estimated savings on healthcare expenditure would be $4.4 billion a year.

Walmart, the country's largest retailer, is also starting in-store clinics. Between CVS, Walgreens, and Walmart, the majority of the country now has much better access to basic healthcare. It's also one-stop shopping. If the nurse says what you need is an over-the-counter medication, it's just a couple of aisles away. If the nurse reminds you that your medication needs to be refilled, the pharmacy counter is right there.

Another reason I'm so excited about mega chains getting into healthcare is because they have always been consumer facing. They understand the customer (in this case, the patient) in terms of buying habits and lifestyle and how to make them happy enough to choose their store over another, how to make them happy enough to spend more money in their store, and how to keep them coming back. The mega chains will help change healthcare by changing the customer experience. They work hard to make customers happy and loyal. This is a radical change from the way most of us experience healthcare. The patient isn't necessarily the customer at a hospital or doctor's office—whoever pays the bill is. In healthcare today, that's insurers. For the mega chains, however, the patient really is the customer, and they know exactly what to do to attract and keep you. When was the last time you signed up for a hospital or clinic rewards program? The mega chains will make all healthcare providers start to care a lot more about the customer (patient) experience.

Innovative ways to make healthcare more accessible include making your electronic medical records (EMRs) easier to use. Right now, different record systems don't always talk to each other. Your records from one hospital may not be accessible to your doctor or to a different hospital. This is a really annoying barrier to efficient care in general. Missing or inaccessible lab results, for instance, often lead to repeating the test unnecessarily. Incomplete access could even be dangerous in an emergency situation. One way around this is PicnicHealth, founded by Noga Leviner. She was inspired to create PicnicHealth after being diagnosed with an autoimmune disease. Struggling to keep track of her own health data made her realize that other people needed help too. Your EMRs are very scattered, and not all of them will be available to you even through patient portals with your healthcare providers. Instead of you trying to track down all your records and get them in one place that you and your healthcare team can access, PicnicHealth aggregates all of your EMR data from all sources for you. It's patient-facing and direct to the consumer. You sign up and give Picnic permission to get your medical records from all your doctors and providers, including scans and lab results. They combine it into one easy-to-read patient dashboard that's accessible just to you as the patient. You can allow others to have access as well, which is great if you're seeing a new doctor or you're in an emergency situation.

Sherpaa is a corporate health service that complements a company health insurance plan. It lets employees have access to their physicians 24/7 through phone and text communication. The employee gets accurate, fast advice and treatment and has fewer doctor visits and out-of-pocket expenses. The company gets lower health insurance premium increases, fewer insurance billing problems, and happier, healthier employees. Over a hundred companies are already using Sherpaa. The monthly fee is just $30 for each employee. The fee pays for itself because Sherpaa solves about 70 percent of all health issues without an in person office visit.

Medicast uses the latest technology to help hospitals and health systems bring back the old-fashioned house call. The logistics of a house call have always been a big bottleneck. How long should you allow for travel time? What's the most efficient route from one patient to the next? How can you access the patient's medical records from the home? Medicast provides a system that optimizes the scheduling and dispatch of remote caregivers in order to maximize utilization and minimize windshield time. It's a way for hospitals to expand their care beyond the walls of their buildings.

We're now seeing lots of great desktop and mobile platforms for quickly diagnosing and treating minor health problems. Zipnosis, for example, lets a health system or hospital offer virtual diagnosis and triage to patients. A great innovation here is that if the system determines you should go to the hospital, it gives you a ZipTicket. This is a QR code sent to your smart phone that acts as a digital boarding pass when you get to the clinic or ER. Your information is all there when you arrive, avoiding delay and duplication.

You can arrange for your own mental healthcare with Breakthrough, which offers online therapy. Finding the right mental health therapist in your geographic area can be a big problem— the right person for you might not be accessible. Three major studies have found that online counseling is just as effective as in person counseling. With Breakthrough, you schedule an online session with a licensed therapist or psychiatrist who's skilled in working with patients like you. You're no longer limited to working only with a therapist in your geographic area and to sessions that can take place only during limited hours.

You can finally get the help you need conveniently, and other studies show that finding the right therapist is the main factor in deciding to continue treatment. Breakthrough saves you money on travel time and expenses and lets you have sessions without missing work. All you need is a computer, webcam, and access to the Internet. You can test the fit between you and a potential therapist by communicating before making an appointment for a session. If you feel comfortable with the potential therapist, take the next step. If not, interview someone else until you find a good match. Breakthrough's proprietary software provides a secure platform for patient privacy; the system is HIPAA compliant.

Neurotrack is developing technology that can diagnose Alzheimer's disease three to six years before the first symptoms occur. It uses a computer-based visual cognitive test that may be able to help physicians predict the onset of the disease. This is valuable because many researchers believe that the best opportunity to treat Alzheimer's is to detect it early before significant neurological changes occur. Early diagnosis could also help lead the way to new therapeutics to slow and eventually prevent Alzheimer's disease.

Amer Haider's company Doctella is a checklist app designed for patients about to have surgery. For whatever procedure you're about to have, the app gives you specialized checklists to help you navigate the hospital stay, suggest questions to ask the doctor, handle the recovery period, and more. It's meant to keep patients and caregivers informed and involved with their own care. You download the app from your hospital or doctor. The checklists now cover more than 90 percent of all medical procedures performed in the US.

The Doctella patient checklists are partially based on cofounder Dr. Peter Pronovost's checklist for surgeons, developed at Johns Hopkins, and Amer's idea of seeing your hospital visit as you would a vacation. Just as you would plan out your vacation, you can plan your hospital stay. On a vacation, you know exactly what it's like to go through the airport, you know how to get your tickets, you know what to wear, you know what to pack, and you know what the hotel experience is going to be like. Why not start looking at going into the hospital the same way? Doctella helps you know everything to expect and when and where to expect it from start to finish. The checklists will remind you of things to do and point out things you might never have thought of. Just about every question you can imagine is already asked and answered on the checklists. It's easy to read, easy to understand, and it's on your smart phone. The passport element lets you do things like add personal preferences and fill out forms before your arrival at the hospital. If you need to see a doctor or go to the emergency room, your data can speak for you and get there before you.

At the government level, in 2010 the Department of Veterans Affairs and the Department of Health and Human Services began collaborating on the Blue Button program. This is a tool to make patient medical records easily available for VA and Medicare patients to download and share with members of their healthcare team. This was a transformative moment. Never before had healthcare consumers had such easy and complete access to their records to use how they wish. Today, that transformative healthcare delivery concept is continuing on with the Argonaut Project, which is designed to enable expanded information sharing for electronic health records and other health information technology. In other words, patients will have unrestricted access to their health data.

Aneesh Chopra, the first Chief Technology Officer of the United States, is leading the Argonaut Project. His goal is for everyone to have unrestricted access to their health data and be able to download it to any app. If you're a Medicare recipient, for example, you'll be able to send your information wherever you want it to go just by clicking on the blue button at Medicare.gov.

Conclusion

TRULY PERSONALIZED MEDICINE

PRECISION MEDICINE + WHEN YOU WANT IT, WHERE YOU want it, and how you want that healthcare administered = Truly Personalized Medicine. If you want robotic surgery in the middle of Africa, your physical therapy by virtual reality, your biomarkers continuously monitored with subcutaneous sensors, video games as treatments, your daily medications dose based on your unique genes, and your microbiome as measured that day, pretested on a 3-D printed section of your liver or kidney, 3-D printed pills for per dose, autonomous drone deliveries of medication or devices. The doctor can beam in over a telepresent robot that is also autonomous and will bring you water, serve as a companion, insert an IV or take blood, be or administer a diagnostic tool, and serve as a caretaker that can inject you with hard working nanorobots. Patient experience and treatment plan fully individualized to a patient's lifestyle, personality, tastes, genetics, environment, and desires.

Personalized and precision medicine will completely change the practice of medicine. It will start at the most fundamental level with your genetic profile. From there, your medical care will be personalized just for you. If you have a chronic condition, your care will be based on your individual case. You'll be able to track your health at a very detailed level. Every aspect of your preventative care, diagnosis, and treatment will be unique and designed for you personally. The days of one-size-fits-all medicine are rapidly receding into the past.

Medicine will be tailored to the individual. Instead of trying out a drug on you to see if it works or if you'll have a bad side effect, you'll know in advance which drug is best for you, and at what dosage, based on your genetic makeup by testing you in advance with very fast and easy lab work. You'll be tracked individually with telepresent medicine. For example, Health-Loop, founded by Dr. Jordan Shlain, will let your doctor check in on you daily. You'll be pinged if something isn't right, like a fever or a drug reaction, and the problem will be caught earlier when it's probably a lot easier to fix.

Personalized medicine means targeting and completely individualizing the diagnosis, prevention, and treatment of disease. All the technology I've discussed in this book is converging more and more, day by day, to make that possible. Personalize medicine means faster, easier access to care, ranging from a lab-on-a-chip at your local cvs to telepresence consultations with specialists thousands of miles away. It's medicine your way, completely tailored to you, accessed the way you want or need it.

Personalized medicine will have major impacts across the entire healthcare system. Let's say you're at home and start feeling really sick at 2 a.m. You're nauseous and have severe pain in your lower right abdomen. What do you do now? You use an app on your smart phone to check your symptoms or a tricorder, or you read the data from the sensors imbedded all over your body. A virtual nurse using AI will talk to you, maybe diagnose you with appendicitis, and urge you to go to the closest ER at once. An autonomous car or ambulance will come pick you up, courtesy of a platform company like Uber or Lyft. When you get to the ER, you'll already be triaged because your data will arrive before you do, and the right physicians or healthcare professionals will be waiting for you with your complete electronic healthcare records already open.

At the hospital, doctors determine your appendix needs to come out. Using laparoscopic surgery or a surgical robot, it's removed very quickly through tiny openings, leaving you with very little pain, little to no scarring, and a quick recovery. If for some reason (say you have internal scarring from previous medical treatment) you need open surgery instead, advanced 3-D imaging will help your surgeon know in advance where the problem areas are and come up with the best surgical approach. An exact replica of whatever part needs surgery can be 3-D printed, and the doctors can either practice or refer to it during the surgery as an extra detailed representation.

Any drugs you need, ranging from pre-op meds to the anesthesia to post-op painkillers, will be selected with your genetic profile in mind. Each pill contains the exact dose you need, made by a 3-D printer on the spot. In the recovery area, you'll be monitored by an AI robot. When you go home within a day or so, a telepresent nurse will check in with you to make sure you're recovering smoothly. Your vitals signs will be monitored with small, inexpensive body sensors, from subcutaneous to epidermal to inside blood vessels, and transmitted to your healthcare team through the cloud. If a problem develops, a telepresence doctor will treat it. If all goes well, you'll be back to your normal life within a week. It's why I prfer to call this Personalized Medicine, instead of Precision Medicine, as it is a tailored treatment for the individual in addition to a truly personalized experience. You'll get healthcare where you want, when you want, how you want, and tailored specifically just for you.

Personalized medicine will also make a huge difference in treating cancer. We can already look at the genetic makeup of some tumors and use that to target the cancer-causing mutation with a drug designed to shut it down. Targeted therapy for cancer works by looking at the genetic event that's driving the cancer and matching the treatment to counteract it. We can already do this with some types of cancer, like leukemia and lung cancer. This is an area where the research is moving ahead very rapidly. In the future, every cancer patient will be genetically tested and will receive targeted treatment that will simply switch off the mutation. Broad-spectrum chemotherapy, which kills off cancer cells, will become a thing of the past.

To discover if you have cancer, you'll have a liquid biopsy—basically a blood or urine test instead of an invasive surgical procedure. Eric Topol, M.D., the director of the Scripps Translational Science Institute in La Jolla, California, says liquid biopsies will soon become the stethoscope for the next two hundred years. Liquid biopsies will help detect and analyze molecular biomarkers in blood and other body fluids. They can be used for early detection of cancer and for monitoring cancer patients. They can also be used to determine what the best treatment will be for you and to detect drug resistance early on before more time is wasted on an ineffective treatment. Imagine a world where you can frequently and easily get screened for cancer, more often than you do cholesterol or white blood count! Take it one step further. Converge this with sensors inside of blood vessels, and you'll be able to be monitored for cancer 24/7 and catch it immediately.

Recent developments in our ability to edit genes using CRISPR/Cas9 technology are really changing the face of personalized medicine. This is transforming tissue engineering with possible results we can't even imagine yet. Right now, researchers are looking at 100,000 cancer cell specimens of the 50 most prevalent cancers, which means about 98 percent of all cancers, to find the driver mutations. They've done about 5,000 so far. When that basic research on all the samples is done, a massive step forward will have been taken. At that point, CRISPR/Cas9 technology will be able to edit the defective genes and find ways to stop them. It's all very dependent on massive amounts of computer power and AI. Bill Gates and other investors have put $120 million into Editas Medicine, a start-up using CRISPR/Cas9 gene editing in an effort to eliminate human diseases in previously impossible ways.

At Memorial Sloan Kettering Cancer Center in New York, researchers are finding new ways to do personalized drug treatment for lung cancer. To find out which of the many available drugs will work best for an individual patient, they're working on a tiny implantable device that will carry small doses of about 30 different drugs. The device is implanted in the tumor where the drugs are released directly where they're needed. The implant is a tiny tube that's only a few centimeters long. It delivers the drug dose molecule by molecule, and it can closely monitor the activity in the tumor. That means the cost of the treatment can really be cut. Plus, there are far fewer side effects for the patient.

What a lot of future personalized and precision medicine comes down to is looking at the human body as a hardware system with a software system. We want to use the hardware to go in and edit the human software—your genes. This will be the biggest thing to happen to personalized medicine because there's nothing more personalized than editing your own genes.

We can now decode your DNA quickly and cheaply. Researchers are now working on decoding your microbiome as well. This is an area generating a lot of interest. Your microbiome is the ecological community of symbiotic and pathogenic microorganisms that literally share your body. They live in you and on you by the trillions. In fact, 1 to 3 percent of your total body mass, as much as three pounds, is actually the microorganisms that inhabit us. They outnumber your human cells by ten to one.

Just as a rain forest is a balanced ecosystem because it's so varied, human beings with their integrated microbiomes are balanced ecosystems as well. The balance naturally shifts and changes and, we're starting to learn, can have a powerful impact on our physical and even mental health. This will be the next big frontier in personalized medicine. Once we understand the microbiome, we can find ways to alter it to solve health issues. Imagine, there is a potential that some diseases we've thought of as genetic could instead be infectious through our microbiomes passed down through generations. Getting a microbiome transplant may treat or cure diseases we thought were autoimmune.

That research is already underway at uBiome under the direction of the amazing cofounder and CEO Jessica Richman. To get your personal microbiome analyzed, you just take some sample swabs at home, using a sample kit, and answer a short survey. The DNA of the bacteria on the swabs is sequenced and analyzed. You get a report that compares your microbiome to those of other groups, say vegans or people who take statin drugs. The information isn't a diagnostic test (yet, though they are developing that). It doesn't predict disease, but it certainly tells you more about yourself. It also lets you experiment. You can compare your microbiome today with past samples to see how any changes in your lifestyle or diet change your bacteria. At uBiome, the long-term goal is to collect enough microbiome samples to advance our understanding in this area. Jessica also imagines a world where you bank parts of your microbiome when you're at your healthiest to transplant back if you are sick. In my case, maybe I'd bank my gut microbiome when my Crohn's disease was really under control to transplant back when I have active disease again.

THE FUTURE OF WELLNESS

In the future, a huge amount of healthcare resources will be needed to treat people who have chronic diseases such as type 2 diabetes and high blood pressure. Treatment needs to go beyond drugs and ten minute checkups every few months. Personalized programs for managing your health have been shown to be the most effective, especially for weight loss and chronic conditions. With the growth of AI, apps for this are now becoming a reality.

Companies like Sean Duffy's Omada Health, a company incubated at Rock Health, offer digital health programs that help people prevent and manage chronic disease through personalized programs that use technology to tackle the problems from every direction. Personal health coaches, computerized scales, personal monitors, support groups, games and incentives, and other approaches are combined for maximum effectiveness. The outcomes are good: the average participant loses ten or more pounds over 16 weeks and keeps them off. More importantly, the behavioral changes Omada encourages can lead to healthcare cost savings of over $2,000 per participant over two years. This is targeted, monitored, integrated, personalized treatment plans with daily feedback and minute-by-minute personalization of the program to fit the individual.

The approach to personal wellness and treating chronic conditions at Arivale is to use cutting edge science to translate your personal data into specific, actionable recommendations that will help you thrive. Arivale, cofounded by the pioneering biologist Lee Hood, combines information from your genome,

176 THE PATIENT AS CEO

from body fluids such as blood and saliva, your gut microbiome, and your lifestyle metrics to help you discover ways to improve your overall health, meet your personal goals, and minimize long-term problems. All the baseline data is given to you and your personal health coach, who provides personal recommendations and helps you track your changes over time.

THE VOICE OF THE PATIENT

When you have a chronic disease, you can't forget that you have the disease. Maybe for a second, maybe for a minute, but, at least in my case, definitely never for a full day. You can't take a vacation from it. It's not the thing that makes you want to get out of bed in the morning. But you can still have purpose in your life and make a difference. For me, focusing on changing medicine through technology to help patients gives a purpose to my life. It's something bigger than dealing with doctor appointments and pain. I don't have a medical degree, yet I've spent my life working in medicine. We need patients on the inside, working in the start-up companies that are changing medicine, working with the pharma companies to create new treatments, working with government, and working with companies large and small. Some of the companies I work for are virtual. I work remotely from anywhere I want, including home, a doctor's waiting room, a hotel, or even a hospital bed. If I'm not feeling well that day, I work lying in bed or on the couch.

If you're a patient, you have a very valuable voice. Your views on how to start to rebuild medicine from the inside out, so that it's better for everybody, need to be heard. You personally can have an impact on medicine, healthcare, biotech, and patient's lives with an abundance of options. You can start with something simple like donating your medical data (even anonymously) by contributing to platforms such as Jamie Heywood's successful patient community PatientsLikeMe or taking surveys at 23andMe to help amass a more complete genetic database. Stanford is conducting a research study by having people download the app MyHeart Counts to record their daily activity on their smart phone and donate the data to be used to study heart disease and stroke.

One of the great things about future medicine is that patients can empower themselves to make a difference. You don't need an advanced degree to agree to share your data with researchers or participate in a clinical trial. If you can't leave the house, you can still participate in peer-to-peer groups and volunteer remotely. Or you could start your own company and become one of the disruptors. When you have something that you know will help people, it motivates you. I have Crohn's disease, but none of the health projects I'm working on now are directly related to curing Crohn's or helping people with IBD. I'm working on medicine as a whole and trying to make an impact for millions and billions of other patients.

IT'S HAPPENING NOW

The point of personalized medicine is it allows all patients to be treated as the unique individuals they are. All the factors, all the technology, and all the innovative companies I've discussed in this book are leading us to very targeted personalized medicine.

We couldn't have our current level of personalized medicine ten years ago. And ten years from now, we'll look back to today and say, "That's what you call personalized medicine? Look what we're doing now." What we're going to see over the next decade is accelerating technology that will radically change how we view our health. What will your healthcare look like in the future? It will be mind-blowingly different, maybe in ways that I didn't talk about in this book because they don't exist yet. It's coming, though, as the technologies converge.

The megatrends and technology that will impact medicine give us a lot to think about. They raise questions about FDA approval, patient privacy, the regulatory environment, and of course, how to pay for it all. A lot of the hurdles going forward aren't in the technology; they're in providing access for the individual patient.

A lot of the emerging medical technologies will help personalize your treatment, but they won't always enable you to take control. You're not going to use a surgical robot on yourself. But the new technologies, and knowing the landscape and what's coming down the technological pipeline, will help you be a better CEO of your health. You'll have more data for decision-making, you'll have a better idea of what your treatment alternatives are, and you'll be able to control some aspects more.

You're the visionary, but you'll still need help with strategy and tactics. You'll be able to find that help much more easily in the future. Just like the CEO of a corporation where you surround yourself with a fantastic array of VPs, advisors, and support staff, they do their jobs, report back to you, and together you determine a direction for the company to go. But as CEO, you're the one who's ultimately responsible that the vision gets carried out and that the company as a whole is successful. Why should being a patient be any different?

Now that you are the CEO of your own healthcare, how are you going to start to change your behavior today?

The future of the patient *is* the future of medicine.